HOTEL ORGANISATION AND THE FRONT OFFICE MANAGEMENT

HOTEL ORGANISATION AND THE FRONT OFFICE MANAGEMENT

A. P. Rastogi
Director
J. P. Institute of Hotel
Management & Catering Technology, Meerut

AMAN PUBILICATIONS
New Delhi-110002

Published by:
Rajiv Jain for Aman Publications
LG-4A, Ganpati Bhawan,
4675-B/21, Ansari Road,
Darya Ganj, New Delhi - 110 002
Ph. : 23255012, 23282127
E-mail: sshjybs@yahoo.com

Reprint 2020
ISBN-81-8204-004-3

Sole Distributor :
JINDAL BOOK SERVICES
LG-4A, Ganpati Bhawan,
4675-B/21, Ansari Road,
Darya Ganj, New Delhi - 110002

Printed by :
Tarun Offset Printers, Delhi, India.

Preface

Everything the guest sees, feels, hears and experiences happens in the front line service areas. So the management of front office operations in hotels, motels, and other lodging facilities present particular challenges to managers. A front office manager has to perform various herculious functions including reservations, guest check in, front desk, guest check out, night audit, housekeeping, property maintenance, general manager, and back office interface.

Service providers look to their senior managers and executives for direction on how to handle the critical customer contact moments occurring every day. They also look to management for clarifications on day-to-day quality performance. When leadership presence is absent, service providers have little insight into what makes up quality performance.

The present book uses a human resources approach to cover the unique management and operational challenges in the front offices of today's hotels and lodging facilities. It emphasises on applying theory and management strategies, as well as providing new material on technology's role in the modernized front office. Important aspects such as lodging operations, front office management, guest services etc., are described in detail. The work will be an invaluable reference book for those interested in learning more about the hotel's front office.

Editor

Contents

1

Essentials of Front Office Leadership

Managers are objective and honest in providing feedback to and information about employees. Perceived mid-manager job competence is high. Performance feedback is continuous and frequent; special efforts made by employees are recognized. The practice points in *Guide For Hospitality Managers* have been developed in 2003. Divisions and departments experiencing significant statistical improvements in their overall performance and employee satisfaction scored higher on the previous four practices compared to divisions and departments with zero to negative overall changes.

PRACTICE 1: LEAD THE FRONT LINE TOWARDS SERVICE EXCELLENCE

Everything the guest sees, feels, hears and experiences happens in the front line service areas. Service providers look to their senior managers and executives for direction on how to handle the critical customer contact moments occurring every day. They also look to management for clarifications on day-to-day quality performance. When leadership presence is absent, service providers have little insight into what makes up quality performance. The result is inconsistent service resulting in dissatisfied guests checking out. And, forget the loyalty ratings!

Practice 2: Keep Them Listening With Objectivity And Honesty

When giving and seeking feedback, keep objectivity and honesty in mind as your point of reference. Sometimes the truth hurts, but with tact and professionalism, your team will value your observations and suggestions. The goal-feedback loop is critical to continuous improvement: it informs teams, departments, divisions and the entire organization regarding what to focus on and how best to achieve their goals. When managers are objective and honest in providing guidance and information. the loop is positive and employees at all levels are motivated to achieve targets.

Practice 3: Talk-The-Talk And Walk-The-Walk

Pardon the cliché, but "You've got to be in it to win it." For line employees, especially those with a lot of guest contact, managers who can role up their sleeves and jump right in when it's crunch time or when an extra hand is needed, significantly boost morale and teamwork. Along the same line, employees whose managers are thought to be "in the loop" inspire confidence in their abilities and are seen as important to the company's success.

Practice 4: Provide Feedback That Is Real And On Track

Good performance in communication is indicative of a team's capability to consistently identify and implement goals. When communication breaks down, relationships begin to disintegrate and guests experience inconsistent service. Feedback ensures healthy links between strategic plans, department objectives and the actual day-to-day service guests receive. Continuous and frequent feedback will keep your team's performance aligned to standards for meeting guest expectations. Recognizing your people for their special efforts is a most effective way of encouraging and spreading the motivation to exceed guest expectations.

More Walk in Business

While many hotels have focused attention in recent years on helping reservation agents convert inquiry calls into bookings, most hotel managers still seem to be overlooking numerous other opportunities their front-desk staff encounters every day. One opportunity is the walk-in inquiry. Depending on your market segment and especially your location, walk-in sales can represent a significant distribution channel. With hotels in North America heading into the busy summer leisure travel season, most of which still happens by automobile, the timing is great for those who are looking to eke out a little more topline revenue this year. Unfortunately, most hotels today do a less than adequate job of selling to walk-ins. Typically, the car pulls up out front and someone gets out; sometimes it's the mom, sometimes it's the dad, sometimes it's the 13-year-old kid who is sent in.

Regardless of who it is, they all ask the same question: "How much are your rooms?" All too often associates simply say "$99 a night," without inviting the inquirer to stay or building rapport by engaging them in small talk, let alone any mention of value-added benefits. As any sales professional knows, the worst mistake you can make in any type of sales is quoting a price without demonstrating value. All to often, this leads to the common response of, "Let me check with my spouse outside and I'll be right back."

When you stop to consider it, the walk-in sales opportunity provides the hotel with some significant advantages over other distribution channels, so in theory, front-desk staffers should be extremely successful at converting walk-in inquires. One advantage we have in fielding walk-in inquires is that the salesperson can visually evaluate guests' needs and wants. Are they dressed as if on a business trip or on vacation? What is their age demographic? Are they traveling alone or with family? What is their level of commitment: in other words, do they park the car and walk in with luggage or do they just run in to find out the price?A second advantage is that the walk-in sales prospect can see the product firsthand and is able to formulate a first impression. (This is a reminder of why it is critical for hotels to maintain curb appeal.)

Here are some tips for training your staff to capture more walk-ins:

— *Connect with the customer.* Fundamental guest-service principles will help you gain a competitive edge in any sales scenario, and especially with walk-ins. Greeting guests as they approach the desk before they greet you, establishing eye contact, smiling and using positive body language will set you apart from competitors.

— *Offer a menu of options.* Depending on your inventory of rooms and packages, you'll want to offer several choices when possible. For example, you might offer a traditional room and an executive-level room, or a nonview and view room; or a room option and a suite. Or perhaps your options include a room-only rate and a "package" option including breakfast and/or dinner. By offering a menu of options, the guest is moved along to a decision-making process based on: "Which accommodation option should I choose?" versus "Should I stay here tonight or keep looking?"

— *Use colorful language to "paint the picture."* Common industry terms like "standard room" and "continental breakfast" do little to entice guests. Instead, help them take psychological possession by fully describing the furnishings, views, special-access lounges, and/or food-and-beverage presentations. Remember to use the knowledge gained from reading the guest to point out features that might provide a specific, value-added benefit. For example, be sure your staff notifies business travelers if your rates include complimentary local calls. Hotels offering specialty rooms or suites should also provide front-desk staff with photos to show prospects.

— *Offer to show the room.* Just as the new General Motors marketing campaign encourages the "24-hour test drive" to get drivers behind the wheel so they can take "psychological possession" of the product, once we get a walk-in to actually "walk into" our accommodations, generally their level of commitment will increase significantly. Ideally, a front-desk person or bell staff member can show the room and along the way give them a quick property tour and provide area information. Even if you are a smaller property and have only one or two people at the desk each shift, consider setting aside one room in each category just to show and providing prospects with a one-time access key to go look at the room themselves.

— Avoid positioning last-sell or higher rated options in the negative. Hotels generally experience the most walk-ins during periods of peak demand, which also is when they tend to have the least options remaining. Typically, the last remaining options for same-day walk-ins are either the highest-rated options, such as suites and executive/club floors, or the least-desirable rooms, such as those with restricted views, are next to the elevator shaft or are over the cocktail lounge. Without proper training, front-desk staffers inadvertently position these options in a negative way by saying, "All we have left tonight are our rooms," making them sound overpriced or simply just not worth it.

In addition to training the front desk to utilize these and other sales techniques, it's also important to measure the results both individually and on a department-wide basis. If your property-management system allows, assign a special source code to walk-ins. Otherwise, create a form to manually record your staff's productivity if necessary. Of course, you'll also want to challenge everyone's competitive spirit by posting the results, celebrating your success at monthly meetings and by implementing an incentive when the goals for the number of walk-ins per week and month are exceeded.

TEAM BUILDING

Here is the type of team building and team management you need to be promoting in your organization. Learn five basic guidelines for success. Have you ever gone to the grocery store to buy candy and noticed that one of the smallest candy bars is labeled "Fun Size?" Wouldn't the real "fun size" be one of the largest candy bars? Ever watch successful sports teams? The ones that consistently win and who are near or at the top of their profession aren't necessarily the ones featuring individual super stars. Instead, they feature team players who look like they not only work hard, but they actually have fun doing it—together. High fives, shouts of encouragement, slaps on the back and smiles and laughter all indicate a team that enjoys what they are doing, and one that sticks together like a family.

This is exactly the type of team building and team management you need to be promoting. While each person on "your" team has individual talents, and strengths and weaknesses, as a manager or "skilled coach," you need to mold these individuals into a "well oiled" cohesive unit. Your alternative—to allow a group of workers to clock in and clock out each day, increase costs due to high turnover, lower morale and standards of service and create poor employee loyalty. In order to become a true team—rather than merely a group that spends eight or more hours a day together—managers need to adhere to five basic guidelines for success:

Allow all members of your team to perform and excel at their job. Encourage individuals to support and want to work with each other. Foster team problem solving—allowing for conflicts as well as for solutions. Reward each individual, as well as the team, for their efforts and accomplishments by recognizing them at an appropriate time and place. Most importantly—keep promises and commitments. Unlike sports, where team goals and accomplishments are almost immediate, business goals and accomplishments, for the most part, take longer. Successful teams, whether in sports or dealing in business, know that achieving team goals are more important than reaching individual goals.

Baseball players, for instance, may better serve their team by sacrificing themselves and bunting another player to the next base rather than "swinging away" with the possibility of hitting into a double play. Likewise, making sure an employee isn't merely "standing around" when extra attention and help is needed during an especially busy project can certainly help the staff eliminate the possibility of poor and untimely service, resulting in a better team effort and increased success rate for all involved. As their manager, knowing what motivates and drives each player on your team as well as showing support for their needs will lead to increased responsiveness and motivation for each team member.

Be sure to keep everyone "in the loop," communicating both progress on goals, objectives and targets, and changes that affect individuals and teams. And keeping your team "in the loop" doesn't stop at merely informing them of the latest company news. Rather, you and your team should benefit from each other's insights, observations and know-how. Completing the loop means managers must allow and encourage problem-solving feedback. This will encourage a personal commitment to alleviating problems or provide help in finding a solution.

The collaboration of team members will lead to support of methods and goals. Everyone must "buy in" to the program, and everyone must want to be involved. Then, let the team know up front that teamwork will positively influence recognition. A good listener will invariably be able to counsel the problems, consider the gripes and console the tragedies that may befall team members from time to time. But, they will also be able to celebrate the victories and share the good fortune that will also appear. In short, managers need to win the trust of team members, and in turn, the trust should be mutual. They must support team members rather than take all the credit for a job well done. Instead, credit must be given where credit is due.

Good employees will make their team leaders look like stars. In turn, a good leader that keeps commitments and schedules an employee's day off as requested, or in some other way keeps a promise will instill loyalty in an individual and increase the overall productivity and positive outlook for the team. Remembering that *trust is timely, recognition is required and praise is priceless*, a positive attitude enhances creativity and teamwork and "rubs off" on everyone. The more that is expected from a situation, the more success will be received. Even when teams encounter obstacles or problems, having a little fun and using a sense of humor to reduce stress will place the advantage back in your court.

The team approach leads to greater job satisfaction for all. In turn, this leads to happy employees who are willing to give it their all to achieve the goals set for them. And, as previously stated, everyone should be able to have a great deal of fun, enjoyment and satisfaction from their job, rather than just a little. Isn't it really more fun to enjoy the big candy bar rather than the smaller, bite-size version, no matter what it is called?

EFFICIENT INTERNET RECRUITER

Today nearly all hospitality companies are using job boards as part of their recruitment strategy. These sites have two primary functions for employers: job postings, to which job seekers can apply directly and a resume database, in which employers can search for candidates. While technology and features may differ slightly between websites, there are clear strategies that companies can utilize which will make their internet recruitment process more manageable and will enable them to consistently attract the best talent. It is important to understand that these two functions will garner two different types of candidates.

Individuals who apply to job postings (or newspaper ads for that matter) are considered "active candidates". These are people who are actively searching for jobs. They are clearly looking to make a move, as they are not satisfied with their current situation. Furthermore, they may or may not be currently employed. Searching a resume database allows the employer to attract a passive candidate. Most likely this candidate is currently employed. He or she might be quite happy in their current position but are willing to hear about potential opportunities.

The higher performers will typically fall under the passive candidate category. However, it is important to note that "A" players can and do apply for jobs. This is especially true in today's market where there are many talented people out of work. Just as you would create a high impact classified ad for a local newspaper, you want to take the same care in crafting your job posting. In fact, the posting allows you much more latitude than a newspaper ad because there is not a finite amount of space.

Job seekers are typically searching fields from a database so you want to ensure your jobs are visible to all candidates who may be a match. Searching resume databases is more time consuming than posting a job but it does significantly increase your chances of finding the high performer. Again, ask your job board operator for tips on how to best search their database. A good approach is to first cast a wide net and then begin to narrow your criteria. Take a look at the list of job titles and note those positions that may be relevant. If you are looking for a Director of Sales, don't just search for people with that title.

Company size and staffing needs dictate the amount of activity on a recruitment site. The usage may be as little as one job posting every couple of months to having human resource personnel searching databases day and night. Whatever your level of involvement, strive to be as efficient as possible in your internet recruiting and you will reap substantial rewards.

WEB-BASED EMPLOYMENT BRANDING

An examination on best practice marketing techniques to attract top talent. The best practice companies of the hospitality industry have long recognized the importance of building a solid employment brand. Not only is it good for the overall image of the company, it is the surest way of attracting and retaining top talent. Those companies who are heralded as the best to work for are progressive in areas of employee training and mentoring, they recognize the importance of the work/life balance, and offer creative compensation programs.

Today, the Internet often becomes the first point of contact for employers and job seekers. It is important for companies to recognize that their online recruitment strategies are a critical component of their employment branding practices. When communicating via the Internet a level of human interaction is lost. Employers need to take greater lengths to connect with that potential employee. Job seekers are consistently saying the same things about what has appealed to them, and what has turned them off. There are several areas where employers need to focus.

Make your corporate website an effective recruitment tool. These days, you'll need more than just a one page listing of your open positions and a fax number on where candidates can send their resume. The "career opportunities" section of your website is your sales pitch to attract top performing candidates. Have your company's mission statement clearly displayed. Give as much information about your corporate culture as possible. Offer testimonials from current employees on what it's like to work for you. Have all your properties posting their available positions on the site, and make sure it is updated regularly. Allow candidates to submit their resume directly through the website, not only for open positions but for future consideration. This will allow you to store resumes and build your own candidate database. The best practice companies, like The Mandarin Oriental Hotel Group have even tied an online skills assessment to their application process.

When posting a job on an online recruitment site make sure you have clear and comprehensive job descriptions. This is not an area in which you want to take shortcuts. I've seen postings for a Director of Sales position where the description read "Needs previous hotels sales experience", and that was it. The high performing job seeker will typically not responds to this type of posting. Firstly, it looks as if the employer can't articulate what the individual needs to do, a scary proposition for a person coming into a new job. Secondly, there isn't enough information for the "A" candidate to decide if they are a fit. A detailed listing of the responsibilities of the position, the organizational structure, and both personal and professional requirements are the minimum required for an effective job posting. This will also limit the influx of unqualified job seekers applying to the position. The posting should also include whether the employer is willing to relocate, or sponsor an international candidate.

The biggest turnoff for a job seeker is the lack of response to an application. It leaves the job seeker wondering if the employer was even reached, this is especially true with the Internet. Employers are often getting a deluge of resumes, and it can get tough to manage. At a minimum, employers should indicate that only job seekers under consideration will be contacted, or to have an automatic reply system set up for applications. Remember to get your marketing team involved in the employment branding campaign.

SERVICE BASED LEADERSHIP

Leadership principles, principles common to all successful leaders regardless of their outward presentation have been listed in many different forms and contexts, but are remarkably similar in their essence. representative list of leadership principles includes:

— Be professionally and technically proficient.

— Respect and show concern for those you

— Set the example.

— Know your capabilities and seek self-improvement.

— Understand your mission.

— Communicate your expectations.

— Instill motivation and morale in those you lead.

— Build teamwork.

— Make sound and timely decisions.

— Develop your employees.

— Ensure tasks are understood, properly supervised, and accomplished.

— Be responsible for your actions.

While different leadership styles are suited to different personalities and situations, the underlying leadership principles seldom vary and are essential to any successful leader. To the extent that you practice and possess these principles, you will be successful. To the extent that you are lacking in these principles, you will be handicapped, or will fail outright, in your endeavors.

Leadership Traits

By conscientiously and consistently applying the foregoing principles, leaders develop character traits that become the foundation for all the skills and techniques employed to achieve success.

— The will to make things happen.

— Willingness to take and stand by an unpopular view.

— The ability to pace oneself and go the distance.

— Loyalty to the company, superiors, and employees.

— Willingness to make decisions based upon the best available information and analysis.

— Dependability and consistency.

— Integrity and truthfulness.

— Fairness.

— Good judgment.

— Willingness to share praise and shoulder blame.

— Professional decorum, understanding the correct time and place for everything.

— Enthusiasm.

Application of Leadership Skills

Leadership is situational. Different situations and people require the application of different sets of skills and techniques. What works for the General will not work for the Politician. The art of leadership, then, is in the ability to size up a given situation and understand how best to address it. While some leaders have an innate sense of what to do, most learn through experience. Unfortunately for first-time supervisors, such experience is in short supply. Some of you are lucky enough to have been mentored by a respected senior leader or you have worked for someone who taught leadership by example. But whatever your examples, much of your success or failure will come from your attitudes about the job and the people you lead.

In extreme cases, new supervisors believe that it is only through their efforts that progress is made, that employees can't be trusted, that goals can only be achieved by driving employees hard, and that discipline is the only way to keep staff in line. Quite naturally, these attitudes create an environment where employees are fearful. Such fear-based management is damaging to your service team and, ultimately, to the company.

Fear-based Management

Fear-based management is rooted in the insecurities of the supervisor. While everyone has insecurities, in this instance, the immature, inexperienced, and untrusting attitude of the supervisor dominates the workplace. Some symptoms of fear-based management are:

— Employees covering their backsides.

— Unwillingness to take a risk.

— Lack of initiative and acceptance of the status quo.

— Employees afraid to express opinions or answer questions.

— Lack of trust.

— Defensiveness and blame placing.

— Lack of communication or only top-down communication.

— Poor motivation and morale.

— Lack of cheerfulness, friendliness, and smiles.

Fear-based management impedes organizational teamwork and effectiveness but can be overcome by a leader with an open, trusting attitude and a willingness to grow as a person and a leader. Because of its detrimental impact on employees, customers, and the bottom line, fear-based management should not be tolerated in any company.

At the other end of the spectrum is the ideal of service-based leadership. With this approach, the attitude and primary motivation of the leader is service to others – to customers, to employees, to shareholders. This approach to leadership naturally creates relationships – the deep and abiding bonds that sustain the efforts of the company. This outward focus of the leader sets up a dynamic where:

— Employees are continually recognized.

— There is an open flow of ideas, opinions, and information.

— Initiative and risk are highly regarded.

— Problem discovery and solution is a focus while placing blame is unimportant.

— Every employee feels energized and part of the team and is valued for his or her contribution.

— Prestige is derived from performance and contribution, not title.

— Customers are treated well because employees are treated well.

— The energy and initiative of all employees is focused on the common effort.

With service-based leadership, you will find that good customer service to both internal and external customers is effortless. Less energy is expended in processing complaints, grievances, and conflicts. Work is more fun and everyone's job is easier.

Leadership in Business

The purpose of any business is to provide a product or service for its customers, make a profit for its owners or shareholders, and maintain itself as a "going concern." In today's highly competitive marketplace, where many companies are vying to provide similar products and services, often the only distinction between enterprises is the level and quality of service provided. Realizing this, it would not be an overstatement to say that the nature of business is service. As a supervisor all your efforts and those of your employees are directed toward providing the highest possible level of service to customers. To do this you must anticipate their needs and desires, set standards of excellence that challenge employees, train continually, and constantly examine and improve the details of your operation.

The success of your efforts and the excellence of your operation are ultimately measured only by customer satisfaction. And it is only through the dedicated efforts of your willing workers that you can consistently achieve customer satisfaction. Success in business is simple in concept, but difficult in execution. In simplest terms, successful businesses give their customers what they want; at a price they think is fair, with the expected level of service. The formula for success can be expressed as:

Right Product + Right Price + Right Service

People are complex, carry lots of personal baggage, are often poor communicators, and are sometimes defensive, quick to lash out and place blame. Often they have agendas hidden even to themselves; they suffer various insecurities and lack clear aims. These people are your customers, your employees, and the owners or shareholders of your company.

As a supervisor your role is to keep all of these people relatively happy and satisfied and, in the case of employees, to keep them focused on the task at hand. The difficulty is that the needs and desires of these constituencies are often in direct competition. Employees want realistic wages and satisfying work. Customers want value and service. Shareholders want to be associated with a high quality enterprise and to maximize the return on their investment. The job of the supervisor is to balance these competing needs. If one gets out of balance, the other two suffer.

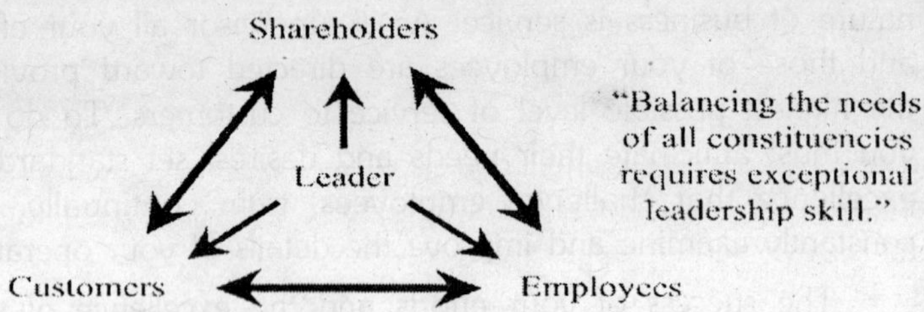

For example, employees are unhappy with poor working conditions and request a new break area. Shareholders don't want to bear the expense. Poor employee morale affects the customers who sense hostility and a less-than-helpful attitude. Customers pick up on the bad vibes and take their business elsewhere. The decrease in sales reduces profit and ultimately the shareholders' return on investment. Everyone is unhappy and blames the leader.

Not only must you be an outstanding leader, capable of maintaining a positive attitude in the face of daily challenges, but you must also remain focused on goals, lead by inspiration and example, and hold employees accountable for results. Further, you must be able to maintain this effort and pace over the long haul. Clearly, this requires men and women of uncommon talent, dedication, drive, and temperament.

Because the quality of leadership is always examined in hindsight when success has been won, we look to the leader with reverence and awe. Often his or her life and accomplishments take on mythical proportions. Thus leadership has become an almost sacred word, wrapped in an aura of mysterious gift and ability. In this light, leadership is always good since it is always applied to the successful. A failed leader is no leader at all. But this is always the historical judgment. For those attempting to lead at this moment, the jury is still out.

Just as a highly competitive basketball game cannot be won without the talent, commitment and effort of all players, so a leader facing a challenge or working toward a particular goal needs the willing commitment of all employees. Without this commitment, the leader is handicapped from the outset and will find success neither complete nor easy to come by. As there are many types of leaders – political, intellectual, military, industrial, business, financial, athletic, and so on for any conceivable endeavor and profession - the skills and abilities that might make a person successful in one arena do not necessarily guarantee success in another.

Many people hold positions of authority by virtue of ownership, heredity, education, oratorical skill, appointment, election, circumstances, or even bluff, but the exercise of authority doesn't necessarily make them leaders. There are various styles of leadership – the most common ones being autocratic, collegial, nurturing, and moral/ethical. There are successful models of each style.

The Freedom of Taking Personal Responsibility

Personal freedom is often thought of as the absence of responsibility. In this respect, no one is free. Everyone is responsible for and to someone else. There is, however, a freedom that comes from accepting personal responsibility for oneself and one's sphere of influence. When you blame no one else for the challenges you face, when you realize that where you stand today is the result of all your past decisions and indecision, you look to the true source of any difficulties. It is never the undefined "they." It is always the ever present "I."

Realizing this is the true source of your freedom. Instead of being buffeted to and fro by uncontrollable forces, you accept the power of your own authority. For good or ill, you are the one in charge of your life. For the supervisor, this means that, as you seek opportunity, you also take responsibility for all aspects of your duties. Size up those around you, your superiors, peers, and employees. If they demonstrate responsibility, learn to depend upon them. If they don't, find ways to compensate for their inadequacies. In the case of your employees, take action as necessary.

In the end, you are the only one responsible for your success or failure. If something goes wrong, there is always more you could have done. In the case of the truly unexpected event, it's not so much what went wrong as how you respond to it. Instead of blaming circumstances or others, take responsibility to make things right. By accepting this degree of personal responsibility, you free yourself from the unpredictability of life and those around you.

Taking personal responsibility equips you to assume a leadership role. But the will to lead is a far cry from being willing to lead. A good number of people are willing to accept positions of leadership. But accepting and exercising leadership are two very different matters. Having the will to lead implies a commitment to face whatever challenges may present themselves. Simply put, it's the will to make things happen.

Owning Your Failures

When you or those you lead fail in any way, don't make excuses. While there may be mitigating circumstances, you must take responsibility for the failure. Whether you didn't plan or train well enough, you didn't devote the proper time or resources to the matter, you didn't establish priorities, or you underestimated the situation, the bottom line is that you failed. But failure can become your most valuable learning tool! Rather than casting about for others to blame, carefully analyze what led to the failure and see what you might have done differently to achieve a positive outcome. This approach accomplishes two very important things:

— You establish your personal responsibility and authority, and

— You analyze and learn from your mistakes.

Leadership is distinguished by leading, not by hanging back in the pack. When you step forward to lead, you risk the chance of highly visible missteps. Remember that experience and "trial and error" can be life's most powerful instructors. Often the greatest lessons are learned from mistakes. Winning breeds a sense of supremacy and complacency; whereas losing encourages critical review. Keep this in mind as you blunder along the way.

There is far more to do every day than you can possibly accomplish. As a supervisor, where you apply your time and energy is of critical importance to your team's efforts and success. Avoid frittering away your personal resources on marginal activities. Focus on the important things that will make a difference in your team's performance. To do this you should make a list of the key items on which to focus and have a long-term plan of improvement for your operation, as well as a list of needed projects to accomplish. Despite the daily distractions and crises that inevitably come up, keep your focus on those key items. When things slow down or windows of time open up, refocus yourself and your team on those important goals. It's also helpful to get away from your operations, even for just a day. Being away from the day-to-day problems will give a broader perspective on the issues you face and will help you recognize where to apply yourself.

Being an effective supervisor requires that you understand the parameters of your authority and the extent of your sphere of influence. For instance, what should you do when you report to a superior who lacks essential leadership skills? Remember the freedom of taking personal responsibility. You cannot control your boss' skill or lack of it, but instead of getting upset, focus on what you can control. Do everything in your power to be the best leader you can be. Try to insulate your team from the worst effects of the situation. Do not disparage your boss in front of your employees. They will size up the situation quickly enough and will respect you even more for not trying to make him or her look bad.

Possibly your efforts will have a positive effect on your boss. If your area of the operation is performing well because of your leadership, it may cause him or her to take notice. Maybe your boss will become curious enough to ask about the secrets of your success. In any case, focus on your own efforts. If the situation should become untenable, remember that you retain ultimate control over your future and can make the appropriate decision at any time.

Understanding the limits of your influence also entails the recognition that you have more impact and control over your employees than you do your other constituencies. Your customers are removed from your direct influence since they are served by your employees. For the most part your influence on customers is secondhand.

Farthest removed from your influence are your company's shareholders. Unless as owners they take a direct role in your company, they are usually absent from the operation. Their role and status with the company is still of major importance, but their interests are served at a distance. So as you work to accomplish your company's goals, concentrate on those nearest at hand and those over whom you have the greatest influence – the employees on your service team. If they do their jobs with enthusiasm and a sense of service, the needs of your other constituencies will also be met.

Serving Your Employees

A supervisor's primary constituency is staff. They are the people over whom the leader has the most influence and control. They are the people who provide service to your customers. In some companies, employees are treated merely as a means to an end – the end of making as much money as possible for the company. Little concern is shown for the employees' work environment, very little time is devoted to training, and employees must often make do under difficult circumstances without the support of their superiors. This should not be the way you treat those who are routinely claimed to be a company's "most important resource."

Your ultimate concern is satisfying the customer, and how your employees are treated has an immediate and direct bearing on how customers are treated. Employees who feel good about themselves, whose welfare and problems are attended to in a supportive way, who are provided with the right tools and training to do their jobs, will continually and enthusiastically communicate their satisfaction to customers in countless small ways.

As a leader, therefore, it is in your best interest to be as supportive of employees as possible. They are the people without whom you could not do your job. If they are unhappy and dissatisfied, they will feel no loyalty to you or the company. You will constantly be dealing with customer complaints about staff and service, continually enduring employee turnover with its attendant training burden, and forever suffering from the turmoil created by unhappy people.

Communicating with Employees

People have a great need for information, particularly when it pertains to them and their environment. In the absence of information, people fear the worst. Employees who are informed about the goals and direction of their company are better able to contribute to its success. As a leader you possess a great deal of information about the company. You attend meetings where goals, problems, projects, and other operational details are discussed. Much of this information should be communicated to employees. Seldom is there a reason to keep them in the dark.

When leaders become absorbed in their own sense of urgency about plans, projects, and priorities, it is easy to forget that employees lack this valuable information. To foster this same sense of urgency in employees, communicate the details of such planning when appropriate. Develop the habit and the systems to constantly communicate with employees. While memos and written material help, periodic departmental meetings allow dialogue and give employees an opportunity to ask questions and express concerns. Every day you talk to one or more of your employees. Use these conversations to paint the vision, explain the big picture, and connect the small daily task to the larger goal.

The extra few minutes spent doing this pays immense dividends in employee commitment. The trust, confidence, rapport, and self esteem engendered by such communication will become a driving force of energy and enthusiasm within your enterprise. Communicating is a complex process. We communicate not only with words, but also with facial expressions, body language, and tone of voice. Words themselves are imprecise and subject to many interpretations.

Some words have different shades of meaning. "Tenacious" has a totally different connotation than "stubborn," and to describe employees as "hard working" is very different from calling them "workaholics." There are also cultural differences, differing educational levels, varied work backgrounds, and diverse life experiences. Listeners filter everything you say through their biases and pre-dispositions, frequently choosing to hear what they want to hear or what they already believe. Finally, communication usually gets garbled as it is passed on.

What all this means is that you cannot take communication for granted. Never assume that someone has understood. It is too easy to be misunderstood. The following points will help improve communications:

— Information disseminated in managers' meetings must be passed to employees. Ensure that information is being passed and understood. Routinely ask employees if they know what information was disseminated at meetings.

— When you make decisions or implement plans, communicate them to everyone directly involved and to other affected areas of the company to prevent things from falling through the cracks.

— Be up front with your employees. Tell them what your expectations and perceptions are and that it's their responsibility to influence your perceptions about their performance.

— Have employees repeat important instructions back to you to ensure they have understood.

— Encourage employee feedback. Communication is a two-way street. It must flow up as well as down. Your employees need to hear from you and you need to hear from them.

— Some communications are so important they must be put in writing. If you communicate in writing, write out the message and sit on it for awhile, then reread it before it's issued. Soft-pedal the tone of written communications, including e-mail messages. Since recipients can't see your face or hear your voice, the message may come across with unintended harshness.

— Documentation fosters continuity. When key people leave a company and take their knowledge with them, the enterprise suffers. Put your policies and procedures in writing so that they are widely understood and passed on properly. Each operational area should have desktop policy and procedure manuals that are periodically updated.

Solving problems

Solving problems is a major function of leaders. Every day new problems crop up and, if not addressed, threaten the smooth functioning of the company. Simply talking about problems doesn't solve them. If so, complaining would be a highly valued professional skill. Leaders must prioritize problems, addressing the more significant ones first. To solve a problem, the causes and underlying issues must be sought out, alternative solutions explored, discussions with other affected departments held, and finally, a *decision must be made*. Even then, the solution is not complete until employees have been informed of the decision and changes have been completely implemented.

Positive Feedback

Since leaders focus on solving problems, it is natural to concentrate on the problem areas of an operation. But in your effort to fix what is wrong don't forget to recognize all that is right. There are seldom opportunities for dramatic heroism in most businesses. However, there are the daily, dedicated efforts of employees faced with monotonous routine, difficult situations inherent in customer service, and detail, detail, detail. Employees should be recognized for the quiet, unprepossessing heroism that this involves.

"Open Door" versus "Open Mind" Policy

Having an "Open Door' policy means being accessible to your employees who have questions and problems. Such a policy will not work, no matter what it is called, if your employees do not perceive a genuine commitment to listening and problem solving. The difference between an "Open Door" and an "Open Mind" policy is the attitude and commitment of the leader. An "Open Mind" policy helps to uncover problems in your operation, defuses potential blow-ups, builds trust, and improves communication. To be successful with an "Open Mind" policy, you must:

— Be available, accessible, and approachable.

— Listen. Let employees speak. Be patient and constructive in listening to concerns and offering help.

— Be fair and consistent.

— Be prepared to get involved and follow through to solve problems.

— Ensure closure on issues.

If you are concerned about being overwhelmed by employees coming forward, set some guidelines such as "by appointment only" or "as time permits." If some employees are abusing an open door, be forthright and tell them to take greater responsibility in solving their own problems.

Two points of caution:

- While making yourself accessible to employees with problems, keep the interests of the company foremost in mind. This means not becoming an enabler for a problem employee or spending too much time assisting one employee to the detriment of the organization.
- Don't get emotionally involved in the affairs of your employees, and don't offer advice regarding their personal lives. Instead, refer them to professionals who are trained to help individuals with personal, emotional, and mental health issues.

2

Information Technology in Guest Coordination

InRoom Connect using in Hotels

Unisys Corporation is wiring hotel rooms for the 21st Century. It's called InRoom Connect — a complete computer system with Internet connectivity, hotel services and extensive local information and entertainment offerings — allowing travelers to leave their laptops at home without losing their connection to the office.

With InRoom Connect, guests have access to these online services in addition to Web-based news, weather, financial information, e-mail, music-on-demand and, eventually, video-on-demand. Guests also experience superior customer service as a result of the system's integration with the hotel's customer relationship management (CRM) tools, which allow hotels to personalize interactions with guests and increase their marketing effectiveness. For example, using InRoom Connect, hotels could pre-load guest preferences such as favorite room service orders and music selections to deliver personalized content and services upon check-in.

InRoom Connect is a complete computer system that allows hotels to "wow" their most frequent customers. Featuring a state-of-the-art PC with an attractive flat-screen monitor, the system allows guests to access hotel services and a high-speed Internet connection whether or not they have a laptop. For those travelers who might need their own computer to access proprietary software, InRoom Connect allows them to plug in their laptop and enjoy the same high-speed connection. The PC- and server-based system enables Unisys to continually add new features to the system, delivering additional services to guests and revenues to hotels.

"Business traveiers ask for e-mail and Internet access more than any other hotel room services," said Michael DiLeva, director of the Hospitality Practice, Unisys e-Business Services. "Unisys InRoom Connect not only offers these basic features, but integrates them as well into a complete in-room package that allows guests to access a variety of hotel and local services and also enhances the hotels' relationships with customers."

InRoom Connect allows hotels to establish or expand upon their brand by serving as the centerpiece of their loyalty offerings. Through its leading expertise in systems integration, Unisys can interface InRoom Connect with existing systems or provide a complete customer relationship management (CRM) system for hotels that currently do not have one. The result allows hotels to brand their properties, cross-promote among locations and build loyalty with frequent guests.

Unlike traditional systems, InRoom Connect's standard in-room PC solution and CRM offerings involve no up-front costs or capital expenditures for hotels. InRoom Connect provides the same type of flexibility as the software application service provider (ASP) model, where users pay for software as they use it instead of having to purchase it initially. With InRoom Connect's services offered to the hotel for a small daily fee per occupied installed room, hotels avoid potentially difficult capital commitments for in-room Internet equipment and CRM services. And with InRoom Connect serving as such a valued-added feature – effectively turning a standard room into a Web-equipped business - class room – hotels are easily able to earn a premium return on their investment in InRoom Connect by charging the guest a daily fee for the PC and CRM offering.

"The Internet is a tremendous tool that can add strategic value within every link of a value or supply chain," said Barbara Babcock, president, Unisys e-Business Services. "Not only will travelers using InRoom Connect have access to services they demand, but they will also receive improved customer services from the hotels through the integrated CRM functions."

In addition, with advances in bandwidth expected to make high-speed Internet connectivity free in the near future, InRoom Connect allows hotels to build a complete value-added system that drives premium room rates to ensure that hotels can replace revenues once generated by long-distance access fees.

High-speed Internet solutions abound, yet it is often difficult to tell one offering from another. Unisys e-@ction InRoom Connect Solution provides the same connectivity, yet reaches far beyond the pack to let you distinguish your hotel from those of your competitors. With InRoom Connect, you can create a personalized environment for your guests and improve management decisions while lowering operations costs.

Unisys InRoom Connect operates via an easy-to-use touch-screen to deliver a new and enhanced guest room experience, including:

- — High speed Internet access, without tying up telephone switches or rewiring
- — Comprehensive and current entertainment on demand
- — Online services, creating a virtual office
- — Information about the local area, conference and hotel services

But that's not all. InRoom Connect provides your hotel with operational advantages, such as:

- — Customized interface to your property management system
- — Expanded database for customer management and marketing efforts

All this is backed by Unisys, a world-renowned solutions provider with over 40 years expertise in the Transportation and

Travel industries. InRoom Connect raises the standard for your guest services, while helping to build brand recognition, customer loyalty and a world-class hotel operations platform.

— Increase revenue and build loyalty with a personalized environment for your guests

— Distinguish your hotel while reducing operations costs

— Improve management decisions and expand conference capabilities

Increase revenue and build loyalty with a personalized environment for your guests Customers drive revenue – and your hotel's success. Their expectations and savvy have forced the industry to respond with increasingly competitive packages and services. Leading hotels now actively search for value-added solutions that meet customer expectations – and capture their patronage. InRoom Connect offers a whole new guest experience, including:

Internet access

Now guests can leave their laptops at home and use InRoom Connect's high speed access to online information: the Internet, e-mail, Microsoft Office products and access to hotel and local information – without tying up hotel switches or missing incoming calls.

Entertainment

In-room movies, music, pay-per-view events, books and games are all easily accessible on demand. And new content is continuously supplied.

Information

A virtual concierge, InRoom Connect gives your guests access to information about local theaters, restaurants, sporting events and other attractions. They can book reservations immediately, and then print maps and explicit directions to the site. Guests can shop at local stores without leaving their rooms. Even last-minute postcards can be bought from the hotel guest shop, customized and mailed on their behalf. In fact, many hotel services can be offered online, saving time and effort for your guest – and giving you new streams of revenue.

Distinguish your hotel while reducing operations costs

In an era of dwindling margins, every enterprise is tasked with streamlining operations. Hotels are no exception. InRoom Connect helps you meet this challenge by introducing an operations platform that simplifies the operations process, while raising the level of guest experience. This unique solution streamlines a variety of critical processes:

- Reduces phone lines and creates a dual phone capability – without rewiring your property
- Interfaces with your property manage-ment system for guest billing and staff work orders
- Replaces time-consuming and costly processes, such as room service menus, guest information guides, mini-bar verification and wake-up calls
- Advertise and promote to target guests
- Build a database to reward frequent guests and generate sales opportuni-ties based on guest preferences

Improving management decisions

Critical management decisions are based on accurate, timely information. And InRoom Connect provides the kinds of statistics and customized reports that help to ensure your decisions are financially sound and targeted. With InRoom Connect you can create reports that impact all of your strategic and tactical processes, allowing you to focus on areas affording the greatest benefits, for example:

- Develop complete audit trails to ensure smoother check-out and accurate billing

— Obtain premium rates from adver-tisers to increase revenue streams·

— Identify guest preferences and concentrate on their priorities

And increasingly, management attention and strategy is aimed at attracting – and retaining – conventions and conferences. These large venues provide almost 30 percent of the Hospitality industry's lodging, and brings with them numerous peripheral revenues. InRoom Connect can enhance your conference offerings as well:

— Display events and ensure instant notification of changes

— Create and maintain personal schedules

— Enable easy messaging among blocked rooms and to convention coordinators

— Provide usage, attendance statistics and reports to conference management

Trouble with Information

From the world's largest blue chip organisations in multiple locations with vast amounts of internal knowledge to the smallest software houses with a single product offering, delivering information that comprehensively and effectively addresses the needs of the audience (be they internal or external) is an objective they all strive to meet.

Most organisations manage a wealth of information including:

— Internal policies, processes and procedures that guide the organisation's employees, support them in performing their roles, and reinforce the culture and business model of the organisation

— Technical documentation that supports the organisation's products and services, and provides a vital point of reference for the organisation's customers

— Marketing and sales related information that defines the organisation's place in the market and supports the sales process at all levels

— Training and development information that is key to nurturing and supporting the performance of the organisation's employees, while enhancing the awareness of suppliers and customers

Not only do most organisations have to contend with these different types of information, they also split the information by department. Consider then the volume of information managed at each level and in each department of the organisation. The challenge of maintaining and distributing this information to all those that need it in a usable, accessible and cost-effective manner is apparent. All too often an organisation's business-critical information is out of date, inconsistent with the information being delivered by other parts of the organisation and crucially does not accurately reflect the requirements of the users.

Some organisations have considerable understanding of the issues they face but are unaware that there are cost-effective (and often fairly easily achievable) measures that can be introduced that make a world of difference. Often they just don't know where to start. Other organisations don't think they have a problem, but continue to suffer inordinate amounts of time and money seeping through the cracks in their information management infrastructure. Consider whether any of the following situations sound familiar:

— Multiple departments create information about the same products, services or initiatives, based on the same basic data but aimed at different audiences and fulfilling different purposes. Not only are these departments duplicating effort, chances are they're all delivering information in a different way, delivering mixed messages, and giving the users (be they internal or external) little commonality or consistency in the information they receive.

— Information is not up to date or accurate. When information is managed disparately, and particularly when it is delivered on paper, there is little chance of ensuring that it is kept current and maintained as an accurate reflection of the product or service it supports. Organisations are put at risk by introducing products or services that are not supported by effective information - even if your user documentation has been updated, are you sure that your training materials and your internal support information (such as your call centre guidance) reflect your new offering? Effective information management can remove this tendency of organisations to 'prioritise', which can often relegate the needs of important audiences.

— Your products and services are market leaders but the supporting documentation doesn't have the 'wow' factor you're looking for. Many large organisations have in place the resources to deliver innovative solutions but rarely do they also have the expertise to support those solutions through innovative information delivery. As a result, your offering may be the hottest thing on the market, but it is supported by the same old style of information (whether it is online or on paper) that you were using three years ago

— The number of customer calls to your help desk or call centre, and the demand for information from within your organisation is an ever-increasing burden. This suggests that the information you are delivering, internally and externally, is either not accessible enough, not effective enough, or both. It is all very well putting in place a new intranet or portal to share information with your employees or customers, but if the information itself is confusing, inappropriately structured or poorly written, then the communication fails. Effective product and service documentation can increase the productivity of your support staff, and reduce time spent on inappropriate activities, such as trouble-shooting

The encouraging thing is that for all of these common problems there are effective approaches and solutions available that not only take the pain away, but also establish intuitive mechanisms and control processes to ensure organisations move forward with their information strategy in the right way. In addition, it may be heartening to discover that these measures don't always include implementing an expensive Content Management System (CMS) or making wholesale changes to existing systems. 'Softer' mechanisms that better reflect the needs of your users can exact substantial business benefits and cost savings, for example:

— Tailored training and support
— Clear information management processes
— Standardised editorial guidelines
— Structured information architectures across all information types
— Delivery mechanisms that reflect better the defined information architecture and user needs

You may be used to technology vendors telling you that by implementing their product your information management and distribution problems will be history. While carefully planned and targeted technology implementation may be part of a solution, the truth of the matter is that technology alone will never solve your problems. The following methods and ideas have been used in delivering solutions to leading organisations, and have brought about tangible business benefits related to information management and delivery.

Integrated approach

By maintaining and creating information in a more inclusive way, where departments work more closely together towards common or sympathetic objectives, an organisation can radically reduce medium-term costs and deliver far more consistent and usable information. InterContinental Hotels Group (IHG), who own a portfolio of well recognised and respected hotel brands including InterContinental Hotels & Resorts, Crowne Plaza Hotels & Resorts, Holiday Inn, and Express by Holiday Inn, have shown that they recognise the value of harmonising their approach to information management and delivery.

By integrating their engineering and brand standards with

risk management information and training materials, IHG are not only streamlining their approach and thereby reducing effort and cost, but providing a far more coherent and accessible set of information to their users. Not only can training and business support information be linked both horizontally and vertically, allowing users to drill down through IHG policy to work instructions, it can be delivered seamlessly to multiple user groups, through harnessing a solid information architecture, and navigation and applying recognised good usability principles.

John Ludlow, VP for Risk Management EMEA, says "In Risk Management it is essential that critical risk assessment information and risk policies and procedures are kept absolutely up to date, accurate and available to the people that need them. We recognised that we could achieve this by maintaining the information online, and making it easily accessible to the various stake holders involved in Risk Management globally, from our hotel managers and security staff, to our auditors and risk assessment specialists. There is a great deal of commonality between the Risk Management related information and the other brand standards, so it quickly became apparent that there was real benefit in moving all of this information online in the same way."

A Modular Information Environment

A modular information environment is one in which the information is split or 'chunked' into individual components dependent on purpose, rather than contained in a traditional 'book' format. In a modular environment the information components are stored in an online repository, they can be searched, checked in and out, automatically tracked through approval workflow and version control, and easily used for a multitude of purposes.

Modular Information Management Model

Such an approach goes a long way towards establishing a single-sourcing environment, where the same information components can be used in marketing documents, training courses, technical documentation and in any other information deliverables. Implementing this approach requires a carefully constructed taxonomy, probably some investment in infrastructure, and possibly a cultural shift, but for large organisations the long-term benefits in cost savings and flexibility can be very substantial. True modularity provides greater control and understanding of your information while simultaneously providing you with an intuitive creation and maintenance model. In this model information can be used and re-purposed for multiple audiences and dynamic product and service variants.

Nokia Networks is one large organisation that has adopted this modular approach in documenting their products and services. With a rich portfolio of products, including multiple versions and variants, it is clear to see where a modular information environment comes into its own. Nokia has been able to leverage the same basic information components for similar products in different markets, in continuing to support multiple variants of the same product concurrently, and for constructing technical documentation and e-learning deliverables quickly and consistently by using information components that already exist and are largely ready to go.

Britannia Building Society, a prominent UK financial savings institution, is one client that has recently made giant strides in the quality and efficiency of its internal procedural information delivery. Prior to the rollout of several new business IT systems, Britannia realised that some effort would have to be undertaken to ensure that suitably targeted information was made available to support these missioncritical systems internally. The processes and procedures existed in a variety of disparate styles, were provided sometimes on paper and at other times in online format, and suffered from multiple contributors and a lack of version control, consistency and purpose.

Although great progress had been made in establishing a new Web site and corporate intranet that would provide staff and customers with a usable and accessible information system, the content itself did not reflect the expectations or requirements of the users. Britannia sought external guidance in developing an information architecture that would be accessible and usable, and that would be a key component in delivering the performance support the organisation desired for its staff. Having defined the architecture, identified gaps and duplication in the information, and determined the exact requirements and objectives, Britannia were able to quickly move forward with creating the information they needed to support the rollout.

Britannia teams were supported in documenting the new systems, and trained in the Information Mapping® methodology; a technique encompassing many best practice information design and writing principles that can easily be taught to non-specialists, enabling them maintain the information themselves in the right way going forward. Adrian Northcott, Project Sponsor for Business Improvement, testifies "All of our staff in branches, in the Contact Centre, and in the Customer Support Centre, now have the information they need at their fingertips to support our customers. We have a great baseline of accurate, usable processes and procedures with which to move forward and expand upon ourselves as we need to".

Techniques and Technologies

Many ideas, techniques and mechanisms for delivering and managing information exist now that simply weren't around three years ago, but many organisations are still employing legacy models for managing their information assets. Innovative approaches are available that have not been widely adopted because the importance of information was often underplayed in favour of elaborate claims surrounding technology. In the dotcom revolution, content was the poor relation of content management and actual learning materials were considered secondary to the learning platform.

The technology-driven dotcom boom meant than many organisations invested in the technology, but the new Web sites, learning management systems and intranets became the dumping ground for inappropriate content. Any leading-edge technical solution can only be successful if the purpose behind the implementation is truly understood and bought into. Organisations must focus on the needs of their users, and target their information accordingly, harnessing the right technology as required, as opposed to making tired content fit technical constraints. Quite often, a fresh approach requiring relatively little investment can make a big impact.

Organisations should look carefully at their existing information strategy, consider usability objectives and perform audience and task analysis that will enable the organisation to put in place the right approach, tailored to the specific needs of their users and reflecting their aspirations and constraints. Technology should be employed as a facilitator, rather than a driver. Organisations that really want to excel in the quality of their information, while ensuring they more effectively address the cost implications of managing ever greater volumes of information should consider solutions like XML migration, document or content management and learning management systems. They should have the aim of ensuring the right information is communicated in the right way, to the right people at the right time.

With the increase in the volume of information that organisations create and manage to support their products and services, and their staff and infrastructure, comes an ever greater need to ensure that the information is coherent, manageable and meaningful. Kudos is the world's leading specialist in information strategy, design and delivery. Kudos is also recognised as a trusted outsourcing partner to many global corporations, including Nokia, Nortel Networks, Fujitsu, Hewlett-Packard and Alcatel. The company is headquartered in the UK, with operations in the US, France, Finland, Germany and Hungary. It doesn't always require a big bang approach, or massive investment in technology, to make a big difference in the quality and cost effectiveness of your information.

Kudos provides business value through delivering information solutions. The core offerings cover information architecture design, content creation and distribution, instructional and e-learning design, content management solutions, Web systems development and globalization. Kudos is recognised as a trusted outsourcing partner to many global corporations. The company has operations in UK, USA, France, Finland and Hungary.

3

Web-based Property Mangement

Web based PMS solutions are definitely growing in popularity as a choice for both chain organizations and independents alike in year 2004. The driving force behind this growing acceptance is the simplicity and reliability of the technology behind them. Ultra thin, web based PMS solutions are lightening fast with property level installation as simple as a software download. With access provided over high speed cable or DSL, the reliability of these web based mission-critical PMS solutions rival the traditional local server approach and offer many additional benefits.

Unlike the traditional local server Windows PMS implementation, these web based alternatives require very little hardware infrastructure at the property level. In most cases, all that is required is a Windows based PC with high speed Internet access (HSIA). The usual complexities of desktop processing speed, printer access, network infrastructure, and server sizing and implementation are no longer a concern or a cost. Hoteliers selecting web based PMS applications are making use of existing equipment and saving big on new hardware purchases.

Web based applications, and the companies that offer them, can provide the hoteliers the traditional comfort of "ownership of their data", without the ongoing IT complexities of managing their own system. Off site hosting is now very inexpensive and makes sense for even the smallest of hotel properties. Hotels can choose their own hosting company and feel comfortable that their data is secure and protected as well as all of the on-going backup, maintenance, database and hardware concerns being expertly managed.

For many of these reasons, Ramesys, an industry leader in the development of PMS solutions for over 20 years, has seen a significant increase in both chain and independent customers preferring a web based implementation of their PMS offerings. According to Rick Kennedy, *"The maturity of the Internet, the reliability of high speed Internet access and the development of our own ultra thin client are all factors that have lead to the trend of our customers choosing web based implementations of our PMS solutions over the traditional local server implementation approach. Using today's readily available technology, we have decreased the traditional IT burden and cost to our customers tremendously and allowed them to refocus themselves on exactly what they have always wanted to do, the management of their own businesses."*

In one such customer example, the Best Western Soldiers Field in Rochester, MN, a 214 room full service, group and Fully Inclusive Tours (FIT) property, installed a 17-user license of the web based Entirety Enterprise Suite from Ramesys, consisting of Front Office, Sales and Catering, Customer Relationship Management (CRM) and web reservation modules. For this particular property, Ramesys made the hosting arrangements on behalf of the customer at a secure, professional off site hosting facility local to their head offices in New Jersey. Property level access and operation of the system has been very reliable and lightening fast.

"As a web based hosted solution, the Entirety Enterprise Suite shifts the responsibility of managing computer hardware away from us and back to the experts. As a result, Soldiers

Field has been able to focus on managing our hotel while Ramesys helps manage our IT," said Myron Salz, general manager and owner at the property.

The multi-property Entirety solution is also one that is well suited and being implemented at the corporate level as a hosted solution for many hotel chains. In addition to the benefits described above for all hosted solutions, as a chain, the benefits would further include:

— A single guest profile across all properties
— Centralized reporting and analysis
— Cross booking capability at the property level
— Centralized web booking (option) across all properties
— Centralized AR (option)
— Centralized direct eMail and guest retention programs.

Web hosted solutions for even mission critical applications like property management solutions are here to stay and growing in number. Old issues of reliability and data ownership are no longer factors in the decision making process with benefits such as simplicity and cost effectiveness playing an even greater role in ensuring that the trend towards web based PMS implementations continues to grow.

The Modern Property Management System

Many new Windows systems were gaining prominence, but the established character-based (DOS, UNIX, OS/400) products were still very much worth considering. A round of consolidation had left fewer but (hopefully) stronger vendors at the top, and a number of foreign vendors (AremisSoft, Hogatex, InnSite, HotelLinx and others) were either looking for U.S. representation or had just become established there. The wider adoption of standards for systems at the hotel – Windows, Protocol Technologies/HITIS, Crystal Reports, MS Office integration – had made it easier to bring new property management systems to market, but external interfaces were still a challenge. Further, increasing systems standardization by the major chains was beginning to reduce the sales opportunities open to the smaller vendors.

Windows Rules

Virtually all PM systems on the market now use graphical user interfaces, either Windows or browser-based, and most of those that don't are moving there. It's simply a more powerful and intuitive way to use a system. The newer applications are built with architecture, languages and techniques that assume from the start that flexibility, expandability and interoperability with other systems are key factors. These days, flexibility of access is as important as flexibility of function. No one is writing new systems for UNIX or IBM iSeries (born AS/400) any more, but these are still reliable and powerful platforms. Vendors with well-established PM systems based on them (such as Galaxy - born Geac - and IAD) are keeping them relevant and competitive with browser-based front ends and tighter interfaces to other applications. UNIX systems in particular have received a boost from the recent popularity of Linux as a powerful and lower-cost alternative server platform.

Playing Well with Others

While vendors continue to expand their PM systems' functionality to cover more and more of a property's operational needs, most must operate with a mixture of systems from other vendors and must exchange data with them. Looking at the great library lists of interfaces offered by PMS vendors in the past, you might think that working with other systems had been resolved a long time ago. However anyone with experience in installing interfaces knows this isn't so. Older systems weren't really designed for data exchange, so for every interface each pair of vendors had to agree on exactly which data elements were to be sent and received, of what length and type and in what order. The slightest change could stop the interface cold, and it's no wonder that inter-vendor relations could become a little acrimonious when a client's interface stopped working.

The arrival of XML – eXtensible Markup Language – has transformed this situation by making each data element within the message self-describing, both for content ("here is a four-digit room number"), and in what functions can be performed with it. As a result, complex data exchanges between different vendors' systems can be accomplished more easily. It may be coincidence, but this seems to have led to a marked increase in the number of vendors working cooperatively together, solving joint clients' problems and leveraging their own areas of expertise into wider markets. The power of partnership is, thankfully, becoming a recognized business plan.

For example, many PM systems can now access the full details of a guest's check in a POS system. You might expect this of Fidelio and MICROS, but it's also true for Northwind's Maestro and Silverware's POS, to name just one other instance. Other PMSs such as Springer- Miller and IAD exchange real-time availability and group booking action item ticklers and much more, with sales and catering systems such as Newmarket's Delphi and Daylight's enterprise. PM systems can now have a potential piece of business evaluated automatically by a revenue management system, with various date and rate options re-displayed directly in the PMS.

Perhaps even more telling in a world where customer relationship management (CRM) is critical, we are seeing property systems exchange guest profile data with other guest-focused systems such as spa management. For example, when a PMS reservation is made in INNfinity or Visual One, they automatically create a matching guest profile in spa systems Rio or ProSalon, allowing spa appointments to be booked at the same time as guestrooms and linked to the guest's all-inclusive travel itinerary.

Agreeing on the Rules

One of the stronger forces affecting this greater interoperability is the Open Travel Alliance (OTA), a group working on standardizing traveler "super-profiles" and data exchange message formats between the airline, hotel and car rental industries. As an example, the OTA traveler record allows each person to have multiple personal preference profiles covering their different needs for airline seats, car size, hotel facilities and room type when traveling on business, for personal pleasure or with their families. OTA lets them describe those in one place and have them available to any travel supplier's system that works with OTA standards.

These don't replace the earlier HITIS interface standards, which defined detailed specifications for data exchange among the systems found at a hotel property. The OTA is more externally focused, and in fact adopted the HITIS specifications as its standard for internal message exchange at the property level. PMS vendors who work to these standards will have a real edge in the more integrated, multi-disciplinary environment the whole travel industry is working toward.

Another initiative to encourage greater interoperability is being promoted by the HTNG (Hotel Technology Next Generation) group. Less data-specific than OTA or HITIS, HTNG works to encourage vendors to commit both to a common set of modern technology standards which simplify data exchange, and to common business ground rules for cooperative partnerships in working with each other. Many vendors have endorsed the initiative and pilot projects are under way.

Distribution Channel Management

Distribution channel management is undoubtedly the hot button of the moment. Every hotel posting different rate and room inventory to the GDSs, such as Sabre and Worldspan, alternate distribution systems (ADS) such as Expedia and Travelocity, on its own Web site and with its reservations agents is using a multi-channel sales strategy, and needs a multi-channel PMS to track it. With Internet booking options breeding faster than rabbits, modern PM systems provide clear reports on the most productive sources, even down to the search-engine keywords used to find the property, and help the hotels compare each against its cost per reservation. None of these channels give their services away, and productivity always has to be balanced against fees and commissions.

But this is really only an extension of the revenue management practices hotels have used for years, making specific rates and room allocations available for different guest categories according to various date and occupancy parameters. Good PM systems have long had extensive and easy-to-work-with rate management functions to control this, and taking it down to the individual channel used for each reservation is just a further refinement, albeit an important one.

It's interesting that even some of today's more straightforward PM systems incorporate surprisingly sophisticated rate and distribution management functions. BookingCenter and U.K.-based RezLynx are two good examples. Even a small property with simple rooms management can benefit from better control of its selling. Of course, a good interface to a separate revenue management system provides even greater control, proactively suggesting changes as the actual booking pace fluctuates. Reflecting this, a number of PM systems have tightly integrated links with revenue management systems such as Opus 2's Topline PROPHET and IDeaS' e-yield.

Among related trends aimed at reducing hotels' reservations costs, some vendors (such as SoftBrands) are developing direct links to the four GDSs, bypassing the Pegasus and Wizcom switches. Other vendors such as Northwind and TimeShareWare provide centrally hosted Web booking engines with direct links to their clients' PM systems. These can simplify security via a central firewall and offer a generally higher standard of system support; the vendors can also use them to provide more leverage when marketing their clients' properties to specific travel groups.

Customer relationship management (CRM)

Customer relationship management (CRM) is an equally

important area, and one at least as subject to over-hyping as channel management. It's just a fancy name for knowing who your customers are when they call, knowing the details of their previous stays (including any problems and their resolution) and tracking and accommodating their preferences. Most PM systems have done this for some time with varying degrees of success, automatically running every new reservation inquiry through their guest history files to look for a match, but duplicates and missed opportunities to shine are still frequent.

Better systems track enough information to be useful and present just enough of it to the reservations and front desk agents to help them without getting in their way; OPERA, Springer-Miller's Host and Visual One are all good examples. The best systems also include tools to identify possible duplicates and merge them into one profile, but even with good equipment it takes dedicated manual efforts to keep your guest profile database clean and accurate. What's different in today's multi-property, data-consolidating and personal service-oriented world is that PMS-generated data is no longer sufficient. Profile and stay data needs to be passed on to a central guest information database, and the PMS must be able to receive profile updates from outside.

Guest preferences are an obvious example, which need to be separated into the property-specific (ocean view) and the guest-specific (feather pillow), but input on a guest's experiences and preferences can come from many places, such as a bellman's or restaurant server's observations. Good PM systems allow this kind of input to be recorded simply and effectively. The other activity usually included under CRM isn't really aimed at taking care of the customer; it's targeted marketing based on analysis of your history database. Certainly you can use this to identify which of your regular guests are likely to respond to a particular new package and send them customized invitations. However, analysis of this database is more often used to identify typical guest demographic profiles, which can then be targeted among the world at large to attract new guests likely to enjoy your facilities.

To support this, a modern PMS needs both a sufficiently detailed guest profile record that tracks pertinent details for a meaningful number of guests, and a powerful, flexible analysis module that helps the marketing team mine it effectively. This is still an area in transition, though. Some vendors include a useful report generator in their basic PMS package, but others still suggest that a hotel acquire and learn an external reporting package such as Crystal Reports or Cognos PowerPlay. The usefulness of the data dictionaries they provide also varies considerably.

Internet booking and access are almost too obvious to mention, but four years ago were a real rarity. With travelers researching their trips and booking online in increasing numbers – Internet reservations are up 500 percent from four years ago – hotels simply have to maintain a Web site, and their PMSs need to be able to take reservations from it. Yes, you can have reservations from your Web site handled by another reservations service before being transferred to your PMS, but if your PMS can handle the bookings directly, why not let it and avoid transaction fees?

Your guests can get the warm feeling of being recognized, too, if you give them a code to access special rates. This works just as well for general discount categories such as AARP as for specific groups like your frequent guest program members or a convention with a confirmed reservation block. Password-protected access lets your frequent guests update their own profile information in your database, and allows group coordinators to enter rooming list names themselves. They feel more empowered and you do less work and have more accurate information; good news all around.

In this Internet-centric world, of course, it's a given that a modern PMS can e-mail reservation confirmations and folios to guests. The latter is useful not only for guests who need a copy for their taxes but also for larger corporations that centralize their travel reimbursement programs. Multi-Systems, Inc. (MSI) handles this chore for corporate clients of Prime Hospitality, extracting folios nightly from their MSI-equipped hotels, consolidating them centrally and transmitting them to the companies involved. As more corporations implement stricter travel policies regarding approved hotel chains, this need may increase steadily.

Rooms Management

Constant improvements are still being made at the property level, both within the traditional guests/rooms management area and in expanding the PMS into other areas. More vendors include sales and catering modules in their PMS (MICROS-Fidelio, MSI, Northwind, Visual One, etc.). Others have added spa management and guest activities management functions; Springer-Miller led the way here some years ago, but Visual One and others now also offer them. And still others offer timeshare management, an advantage since that's one of the few bright spots in the hospitality world right now.

More properties are going mixed-use, including vacation interval ownership units and even private clubs at the same location as transient guestrooms. A number of PMS products cover many of these areas, but it's tricky to get it right given the wide range of reservations rules, entitlements and vacation points associated with the different types of guests. Careful research is very necessary to make sure that a product will actually match your requirements in the depth you need, or that the vendor is willing to make it do so.

Smaller properties have often felt that they're forced to buy more functionality than they need, especially when optional modules tend to be an all or nothing deal. Two more flexible approaches come from the AlohaPMS product (formerly Fabco's LodgeMaster) and MICROS' OPERA Xpress. The former provides good, basic functionality across a wide range of the most common hotel operations, with several options for increased depth of features in areas such as group reservations, housekeeping or accounts receivable.

OPERA Xpress takes the opposite approach, starting with the full OPERA functionality (and software) but letting the property choose any 40 features to be made active from a list of 82, at quite a fine level (do you want to track turn-aways, use wait lists, etc.). This is flexible for the user and simple for the vendor to support.

Multi-property Systems and ASPs

Multi-property systems have become much more common. In the past one hotel might support several others in the same area, sharing central equipment and support services over a dedicated network. These days the server is more often at a corporate office or at the vendor's business, and hotels are more likely to link to it over the Internet using a browser. Both arrangements are included under the same application service provider (ASP) label, although this was originally reserved for pure browser-based systems operated by a third party for a monthly or transaction-based fee.

Either way, they offer many advantages to management groups wanting a more complete view of their properties' statistics and a more accurate, comprehensive guest profile database. Browser-based systems started as fairly simple applications for limited-service hotels, for which they have considerable attractions given their typically lean staffing levels. Now they've become much more comprehensive, with PegasusCentral leading the way.

Remote-access client-server systems have also expanded their range; Wyndham, for example, is implementing MICROS' OPERA across all its properties, hosted from its data center in Dallas, and MICROS itself hosts other chains' operations both in the United States and in Europe. The concept is not for everyone, though, and needs to be approached deliberately with a full appreciation for the costs, savings, risks and advantages. Many properties will always be uncomfortable with the concept of their critical guest and operational data being housed somewhere other than at their own site.

But real-world experience is growing in just how reliable, accessible and secure ASPs can be, and how easily hotel data can be transferred back if the property decides to change systems or is forced to by a change of franchise flag. One other bonus of ASPs and Internet access has been the availability of e-learning. E-learning is Internet-based instruction to teach the hotel staff how to use the software before a vendor representative comes onsite to connect the interfaces and support the cutover to live operations. Apart from providing more flexibility in the training schedule, this cuts down the vendor's implementation cost considerably.

Serving Shareholders

Who are Shareholders? Whether a company is privately or publicly held, the shareholders are the owners of the enterprise. They have invested their money in the venture with the expectation of receiving a return on their investment. Without shareholders and their willingness to assume risks, you and your team would not have a job. Often a company's shareholders are removed from the operation. They allow professional managers to run the day-to-day business of the company.

But even if they are seldom or never on the premises, this does not mean that they do not have needs to be met. While you may have little opportunity to directly serve shareholders, everything you do to serve the company's customers and everything you do to better organize and operate the enterprise does have a direct bearing on satisfying your shareholders.

Shareholders have a number of expectations for the performance of their business. If you do a good job taking care of your employees and customers, it will be easier to satisfy the needs of your shareholders. These needs fall into several broad categories:

Pride of ownership: Shareholders derive prestige and satisfaction from the success of their company and that success is directly attributable to the quality and efficiency of operations. Your leadership has a direct bearing on this quality and efficiency.

Return on Investment: Capital is the dynamo that drives a business. It takes money to renovate a restaurant, to launch a new product line or service, to purchase inventories, to hire new employees, and to expand operations. Shareholders expect a return on their investment of capital.

Asset protection: Shareholders expect that the leaders of the company will safeguard their assets. This includes cash, invested funds, change and petty cash funds, inventories, furniture, fixtures, equipment, facilities, vehicles, and the name, goodwill, and reputation of the company. Leaders cannot fully safeguard a company's assets without the willing assistance and commitment of their employees.

Professionally-*operated "going concern"*: A primary responsibility to shareholders is to operate the business responsibly and professionally as a "going concern," that is, as a business that intends to stay in business.

Every business has a budget, which is its financial plan for the coming period – usually a year. As a supervisor you are responsible for meeting your portion of the budget. The two major components of your budget are meeting or exceeding revenue expectations and controlling your costs. As a leader you should be involved in the budgeting process and monitor actual performance versus the budget. Ongoing failure to meet budget must be corrected by bold intervention.

In restaurants floor manager is responsible for, used a chalkboard in the service pantry to set daily goals for the number of meals served, the average check, alcoholic beverage sales, bottles of wine sold, and other daily revenue challenges. He also posted the month-to-date actual results compared to the budgeted numbers. It was incredible to see how this simple technique galvanized the servers into action. Every day they checked their progress and there was a continual hum of interest and excitement about their efforts to beat "the numbers." Next, the floor manager tied some simple awards into the process as further incentive.

Controlling Costs

Payroll cost is the single largest expense in most service organizations. Be aware of your payroll hours on a daily basis. Keep close control of overtime. Staff schedules must reflect forecasted levels of business. There is a natural tendency to overstaff for the expected peaks of business. Because payroll is such a large part of your expense structure, plan to fill in during busy times, thereby maintaining service while controlling payroll costs. In order to do this you must analyze the flow of your business to determine when you should be available to step in.

By routinely checking your operations during traditionally busy periods, you can readily determine if your help is needed. When you are both willing and able to step in, you build camaraderie, morale, and loyalty among your employees; help control labor costs; and ensure consistent service to customers. Another major cost that must be controlled is consumable items. This can be done by tracking usage rates for key items. Before you can do this, you have to identify items, organize storage areas, and specify stocking levels. A high cost of goods sold may indicate a number of problems, but unless you validate your pricing periodically, there is a good chance that it is directly related to inappropriate pricing.

Because of the volatility of ingredients, menu prices, and portion sizes should be reviewed on an ongoing basis to assure the profit margin on each sale. While menu items must be pre-costed and periodically verified, buffets and receptions should be post-costed on a periodic basis to ensure profitability. Sales mixes must be tracked and operating statistics benchmarked. Be familiar with your costs and look for ways to reduce them. Familiarity with these numbers must be routine and ongoing. Large inventories may ensure that you never run out of stock, but they also lead to spoilage and pilferage while tying up capital that could be put to better use.

Financial and Statistical Analysis

Learn to read financial statements. Budgets should be based upon well-defined assumptions. Compare actual performance each month to the budget and analyze all significant variances.

Benchmark your operation: Benchmarking is the act of establishing a standard of performance based upon existing operating conditions. Benchmarks are derived from measuring and tracking various operating statistics. Operating statistics can include anything you can measure – revenues, expenses, production processes, etc. Examples might be:

— Revenue statistics – measures of the volume of business such as meals served, rounds of golf, number of sales transactions. Also, the average customer expenditure such as average meal check, green fees per round, and average sale.

— Expense statistics – measures of labor cost such as payroll hours and average hourly wage. Also, measures of consumable supplies usage such as napkins used per meal served or laundry chemicals used per occupied guestroom.

— Production processes – measures of the speed, efficiency, or quality of procedures such as guest rooms cleaned per housekeeping payroll hour, inspection scores measuring the cleanliness of guest rooms, etc.

Benchmarks are significant in that they allow you to compare all future operations against the baseline standard. It also allows you to measure one operation against other similar operations or against the industry as a whole. Further, benchmarking allows you to spot emerging trends so you can make timely interventions. Be meticulous and consistent in tracking operating statistics and be up-to-date and conversant with your numbers at all times. The time you spend tracking and analyzing your numbers of even incidental items will make you more knowledgeable about your operation.

How Best to Serve Shareholders

While all of the skills and initiatives are the responsibility of supervisors, they can also belong to your service team. To the extent that you make your employees a part of your efforts, teach them, and share your goals and challenges with them, they will become willing helpers in your quest for quality and service. Imagine the benefit of ten people trying to meet goals and budgets instead of one.

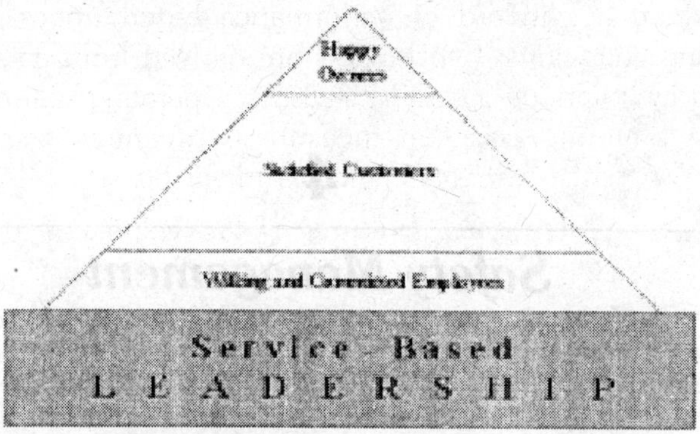

The Pyramid of Successful Service

Think of the natural enthusiasm that can be mustered when your benchmarks become theirs. Recognize the pride of accomplishment they will feel when the team not only meets, but exceeds expectations. This is truly a formula for success. In simplest terms, when you serve your employees, they will serve your customers, who by their continued enthusiastic patronage will serve the needs of your shareholders. We can see the Pyramid of Successful Service depicts these relationships.

4

Safety Management

Fire Safety Management Policy

A fire safety management policy is required, in the first instance, to minimise the risk of fires occurring and secondly, should a fire occur, to enable guests and other persons on the premises to be aware of the danger and to evacuate the premises safely. To implement a fire safety management policy, it is necessary to devise and put in place a practical programme incorporating a number of essential elements. However, such a programme requires personal commitment if it is to be effective. In guest accommodation, the person who runs the business should oversee this programme, or alternatively responsibility could be delegated to a specifically nominated member of staff.

Many of the elements in the fire safety programme relate to good housekeeping and can generally be implemented without significant cost implications. Nevertheless they form an essential component of fire safety management.

Fire Safety Programme

The fire safety programme for guest accommodation should incorporate the following components:

— Fire prevention measures
— Action to be taken in the event of fire
— Ensuring the availability of escape routes
— Maintenance of fire protection equipment and building services
— Keeping fire safety records.

Fire Prevention Measures

Fire prevention measures help to reduce or eliminate the risk of fires occurring. These are essentially items that relate to good housekeeping and awareness of potential fire hazards. Guests also need to be aware of their role in fire prevention. The potential causes of fires include the following:

— Improper use of cooking facilities
— The use of defective heating appliances
— Unsafe use of open fires
— Inadequate supervision of laundry equipment, especially dryers
— Defective/overloaded electrical installations or equipment
— Misuse of electrical equipment
— Defective gas installations
— Misuse of smoking materials
— Improper storage or disposal of waste materials
— Damaged or improper upholstered furniture

This list is not exhaustive, or in any particular order of risk potential. An outbreak of fire could occur for any one, or combination of, these factors. The risks associated with these items should be assessed, and measures put in place to minimise the risk in so far as is reasonably practicable. This exercise should include being satisfied that appropriate standards have been employed in the construction of the premises and on ongoing maintenance. It should also involve guests being aware of their responsibilities and this can be facilitated with the aid of notices, in addition to overall supervision. Fire prevention does not need to be intrusive, but to be effective it requires ongoing awareness and commitment on the part of the person in charge and any persons employed on the premises.

Action to be taken in the Event of Fire

The fire prevention measures referred to above should minimise the risk of fire. However the possibility of a fire at any time for whatever reason cannot be totally eliminated. The greatest danger arises if a fire occurs at night, when people are asleep. The volumes of smoke produced by even a small fire can fill a room or corridor very quickly. The presence of a functioning fire alarm system should provide early warning of a fire. A notice of the procedures to be followed by guests in the event of a fire should be provided in all guest bedrooms, displayed on the back of each bedroom door. This should concentrate on evacuation of the premises when hearing the fire alarm or other warning and drawing attention to the means of escape. Instructions should be clear and concise and should preferably be multi-lingual to cater for foreign guests. As guests are unlikely to be familiar with the internal layout of the premises, the instructions should include a simple location map, indicating the escape routes relative to each room.

Generally, guests should be able to evacuate the premises themselves. However it is important to ensure that they have responded to the alarm and have evacuated the premises. Prompt action is required in these circumstances. The situation is all the more critical where a fire occurs at night, when people are asleep. A fire drill is a rehearsal of a real fire scenario. It is not essential to involve guests in such an exercise and it is not recommended to have an unannounced fire drill, because of the risk of injury.

A drill is an important exercise and will reassure those in charge that they are prepared, if a fire was to occur in reality.

It should involve any staff employed and should try to be as realistic as is practicable. Drills should be carried out at regular intervals and at least at the start of the tourist season and a record kept in the fire safety register. Each drill should be reviewed to identify any shortcomings in the procedures and the procedure should be revised, if considered necessary. If it is safe to do so, it may be possible to extinguish a very small fire, with the aid of a fire extinguisher or fire blanket. It should be noted that portable fire-fighting equipment of this type requires training to use it properly and safely. The person in control and any staff employed should be familiar with the use of any such equipment provided.

Escape Routes

In the event of a fire, it is essential that the escape routes from the premises are available for use and are not obstructed. The following precautions should be taken in relation to all escape routes :

- All escape routes are clearly indicated, are not obstructed, and are available for use at all times;
- The exit doors are capable of being readily and easily opened at all times;
- The external areas at or near exits are not obstructed; and
- The security arrangements for the premises do not impede or prevent the use of escape routes.

Exit doors from the building should be capable of being opened from the inside without the use of a key.

Fire-Fighting Equipment

Fire extinguishers provide a means of extinguishing a fire at the early stages of development, before it has grown to a stage where the occupants are threatened. Every storey of the premises should be provided with a minimum of one nine litres water type or one four kilograms general-purpose powder fire extinguisher (or a number and type of extinguishers with an equivalent rating).

Maintenance of Fire Protection Equipment and Building Services

The person in charge should be familiar with the fire safety features of the premises and any fixed fire safety equipment provided. The principal equipment of this type is that which constitutes the fire alarm system. Non-fixed equipment would include portable fire extinguishers and fire-blankets. The principal building services are the electrical installation and the heating and hot water systems. Maintenance of all these systems and equipment is an essential element of the fire safety programme. All fire protection equipment provided should be in good working order and any staff employed should be familiar with its use. Visual inspection to see that equipment is in place and is ready to use should be done routinely.

Fire alarm systems should be tested regularly and maintained to ensure correct operation. Mains operated self-contained alarms should be tested at least monthly by the use of the test button provided on the units, to ensure operation of the sounders. All alarms should be tested at least once a year to ensure that they respond to fire. Other fire alarm systems should be inspected, tested and maintained in accordance with the recommendations contained in I.S. 3218 : 1989 : Code of Practice for Fire Detection and Alarm Systems for Buildings - System Design, Installation and Servicing. The person in charge can carry out the inspection and testing of the system, but arrangements for the system to be maintained should be made with a competent installation company. Guidance on maintenance requirements for fire extinguishers is contained in I.S. 291 : 1986 : The Use, Siting, Inspection and Maintenance of Portable Fire Extinguishers.

Details relating to the number, type, location and maintenance record of fire protection equipment should be recorded in the fire safety register for the premises. Building services also need to be maintained. Arrangements should be made with competent service companies for periodic checking of electrical installation and appliances, gas installations and appliances and the central heating system. A record of these checks, as well as any maintenance work carried out, should be recorded in the fire safety register. It should be noted that where a material alteration of the premises is to be carried out, these works are subject to controls under the Building Control Act, and require a fire safety certificate. However minor works associated with normal repairs, decoration, maintenance, etc. are exempt from this requirement.

Fire Safety Records

The keeping of fire safety records is an important element of the fire safety programme and for this purpose, a fire safety register for the premises should be established and maintained. The register should contain a complete record of all fire safety matters on the premises, it should be kept up to date and available for inspection if required. The fire safety register should include at least the following information:

— Details of premises, including the maximum number of guests accommodated and details of escape routes;
— Emergency and evacuation procedures;
— A record of evacuation/fire drills carried out;
— Details of fire safety training provided;
— Details of fire-fighting equipment;
— Details of fire alarm system and maintenance records;
— A schedule of all fire resisting doors in the premises;
— Details of emergency lighting and maintenance records; and
— Details and maintenance records of building services.

Premises - Fire Safety

The principal of fire safety measures required for all existing buildings used as guest accommodation, for the purpose of complying with the general duty of care under the Fire Services Act of 1981. Some buildings used for guest accommodation were designed specifically as guest accommodation and should incorporate the fire safety provisions outlined in this part. Others may have been private dwelling houses, or in some other use. Any such change of use is also subject to fire safety requirements and controls, under the Building Control Act. Because of the range of premises that are used as guest accommodation, a guide such as this can only describe the fire safety measures that are necessary for some of the more general situations. By the nature of the issues covered, it is necessary to refer in a number of instances to some technical terms and standards, which may require further explanation and interpretation. An assessment of the adequacy of the fire safety features of individual premises should be undertaken.

Principal Fire Safety Measures

The fire safety measures required for a building to be used for guest accommodation are principally related to ensuring that should a fire occur it will be detected at an early stage and adequate means of escape are provided for safe evacuation of the premises. If an outbreak of fire occurs, its spread and development should be limited by the construction elements of the building such as walls, floors and corridors, until all the occupants have evacuated safely and the emergency services have been alerted. To reduce the risk of accidental fires caused by faulty building services (such as electrical installation, gas services and heating systems), it is important that these services have been installed to correct standards and have been adequately maintained.

Means of Escape

Adequate means of escape should be provided from all guest accommodation. The term "means of escape" applies to the physical means, such as corridors or stairways, whereby a safe route or routes are provided for persons to evacuate a building in the event of fire, to a place of safety. A place of safety is an unconfined space in the open air at ground level, at a point clear of the building where the occupants are no longer in danger from a fire in the building. As a general requirement, a safe route(s) should be provided to enable the occupants of any part of a building to leave by their own unaided efforts in an emergency. The escape routes should be available to the occupants for as long as is necessary to ensure safe evacuation. This will require parts of the escape routes, such as stairways, bedroom corridors, etc. to be protected from fire and its effects. The protection of escape routes includes the provision of a fire alarm system and emergency lighting as appropriate for the particular premises. Consideration should be given to accommodating and providing for the safe evacuation of people with disabilities.

Security arrangements should not be such as to prevent the escape of persons from the building in the event of an outbreak of fire. In particular, fittings on windows and doors should be readily opened from the inside without the use of a key. The extent of the means of escape requirements for guest accommodation depends on the size and complexity of the premises and on the number of occupants accommodated therein.

Components of means of escape

The means of escape provisions consist of horizontal escape routes (corridors) and, in the case of upper storeys, the vertical escape routes (stairways). The horizontal escape routes are those parts of the escape from any part of the building, to a stairway in the case of upper storeys, or to an exit directly from the building to a place of safety. The vertical escape routes are the stairways that lead from an upper storey to a place of safety outside the building at ground level. Escape routes should meet certain criteria in relation to travel distance, availability of alternative escape routes, protection from fire, etc., and these are outlined in the following sections. Parts of the horizontal and vertical escape routes may, depending on the size and complexity of the building, need to be protected from the effects of fire by means of fire resisting elements of construction. In all cases, stairways need to be protected from a fire in habitable rooms opening onto the stairway. In some situations, the stairway may need the added protection of being fully enclosed and separated from the accommodation by means of protected corridors or lobbies.

Alternative Escape Routes

As a general principle, there should be alternative escape routes from any point in a premises used for guest accommodation. The main escape route from the building is normally through the front entrance door, but other routes on the ground floor are generally available by way of doors or windows from other rooms directly to the outside. An escape route from an upper storey includes the stairways, which constitute the vertical component of the escape route. Alternative escape routes from upper storeys require more than one escape stairway and these are required for premises which do not meet the criteria for buildings which are permitted to be served by a single stairway. Alternative escape routes are also required where the distance to be travelled along the escape route exceeds specified limits for specific locations and situations. Where required, alternative escape routes should be remote from and independent of each other. On an upper storey, it should not be necessary to go through an enclosed escape stairway, which could be blocked with smoke, to reach the alternative stairway.

Travel distance

Travel distance is the actual distance to be travelled by a person along an escape route, from the furthest point in the room, corridor or other area under consideration. The total travel distance is the sum of the distances of the different horizontal components of the means of escape.

Certain parts of the horizontal and vertical escape routes from a building need the added protection of fire resisting construction: This principally relates to corridors and to escape stairways, and applies to the enclosing construction, including the walls and doors. There are specific requirements in this regard related to fire doors. Fire resisting construction is generally specified by reference to a time (in minutes) in a standard fire resistance test (e.g. when tested to BS 476 : Fire tests on building materials and structures : Parts 20 to 24). Guidance on the fire resistance performance of other elements of construction is contained in Technical Guidance Document B (Fire Safety) to the Building Regulations, 1997.

Fire doors

Fire resisting doors (referred to as fire doors) play an essential role in protecting the means of escape. A fire door refers to the entire door assembly, including the frame, leaf(s), hardware and seals between the leaf and frame. Fire doors are described in terms of their fire resistance in minutes (integrity test), e.g. FD30S. In this example 30 minutes is the period of fire resistance and the suffix (S) refers to an element of smoke leakage protection. New construction should have fire doors that meet the required fire resistance, and should have certification to this effect. In the case of existing buildings, an assessment of the fire resistance of existing doors may be required, where these have not been installed specifically as fire doors. Many existing doors will have some degree of fire resistance or are capable of being upgraded to achieve a satisfactory resistance. This may involve the fitting of seals and selfclosing devices and additional treatment of critical parts of the door and frame. The techniques for upgrading are well established and may be a preferable option to replacement, especially in the case of heritage buildings where preservation of original doors is an important conservation issue. To be effective, fire doors should normally be kept closed. For this purpose, all fire doors should be fitted with self-closing devices that are capable of closing the door automatically. Fire doors should never be wedged or held in the open position, as this could seriously jeopardise the safety of the occupants in the event of a fire.

Single storey building

In a single storey building the means of escape consists of only a horizontal component. The main considerations in this situation are the limitation on travel distance and the protection of the escape route from a bedroom door to a final exit. For the purpose of protecting the escape route from the bedrooms, the entrance hallway and any bedroom corridor should be separated from all habitable rooms by 30 minutes fire resisting construction. This requirement applies also to the doors, which should have a 30 minutes fire resistance (FD30S), except for bedroom doors where a rating of 20 minutes (FD20S) is acceptable.

In larger premises, corridors more than 12 m long connecting alternative escape routes should be effectively sub-divided by fire doors at approximately the mid-point between stairways. Such doors may be held in the open position by means of an electro-magnetic device that will release the door automatically on activation of an adjacent smoke detector on the fire alarm system.

A protected escape stairway is a stairway that is protected from the effects of fire in the adjacent accommodation by means of fire resisting construction. The walls, doors and ceiling forming the enclosure to a protected escape stairway should have a fire resistance of not less than 30 minutes. It should be noted that in the case of a two-storey private dwelling house, the stairway is not normally constructed as a protected stairway. However, for guest accommodation an escape stairway must have a defined level of fire protection. In a two-storey building with a single stairway rooms may open directly into the stairway enclosure, provided the enclosing construction and doors are fire resisting. In buildings with three or more storeys (with an exception for a small three-storey building) stairways should be separated from the accommodation by means of protected corridors or lobbies.

Ideally, there should be no storage cupboards, etc. within a protected stairway and any such areas should be separated from the stairway by 30 minutes fire resisting construction. Every protected escape stairway should lead directly to a place of safety outside the building at the ground floor level. Escape stairways should also be provided with openable windows, or a roof-light at the top, for the purpose of providing smoke ventilation.

In the case of a building with three or more storeys above ground level, habitable rooms should not open directly into a protected stairway, and should be separated from the protected stairway by means of a protected lobby or corridor at all storey levels, except the topmost storey. In these situations there should be a fire door between the protected stairway and the protected corridor or lobby, i.e. there are two fire doors between any habitable room and the stairway. An exception to the requirement to provide a protected corridor or lobby to the stairway may be made in the case of a building with a single protected escape stairway in the following situations only:

— The building has not more than three storeys above the ground level (i.e. ground storey and not more than two upper storeys), and
— Each upper storey has not more than four bedrooms.

The rooms open directly into the protected stairway. Protection to the stairway should be provided by fire resisting construction including self-closing fire doors to all habitable rooms at the ground, first and second storey levels.

Building with a single protected escape stairway

Alternative escape routes, by way of alternative and independent protected escape stairways, are required where the travel distance exceeds the relevant limitation for travel distance. A single protected escape stairway complying with the requirements.

— The floor of the highest storey is not more than 10 m above ground level; and

— The limitations on travel distance for single escape routes are satisfied.

External escape stairways

An existing external escape stairway is acceptable only where the following conditions are satisfied:

— The floor of the highest storey served by the stairway is not more than 10 m above ground level,

— All doors giving access to the stairway are fire doors, unless it is at the head of the stairway, leading downwards;

— Any part of the external walls within 1.8 m of, and vertically below, the flights and landings of a stairway leading downwards is fire-resisting (note: any windows should be fire resisting and fixed shut);

— The stairway is constructed from non-combustible materials; and

— Protection is provided, by means of fire resisting construction, for any part of the building within 3 m of the escape route, from the foot of the stairway to a place of safety.

Guidance on the construction of external escape stairways is contained in Technical Guidance Document K (Stairways, Ladders, Ramps and Guards) to the Building Regulations, 1997.

Windows

Windows do not form part of the primary means of escape. However they may serve a useful purpose as a secondary escape provision, or for rescue. New and replacement windows to habitable rooms should be suitable for escape or rescue. For this purpose the opening section of the window should be approximately 850 mm by 500 mm and the bottom of the opening section should be between 800 mm and 1,100 mm above the floor.

The evacuation of persons with disabilities, such as impaired mobility, sight or hearing, requires special consideration. It would be prudent to accommodate such persons on the ground floor storey of the premises. General guidance on the provision of means of escape for disabled persons is contained in BS 5588 : Part 8 : 1999 : Code of practice for means of escape for disabled people. The principles outlined in this standard are based on the provision of refuge areas and the management of evacuation.

Where the guest accommodation is ancillary to some other use, such as a shop or public house, the guest accommodation should be provided with independent means of escape. Any connection between the other use and the means of escape from the accommodation should only be by way of a protected lobby.

Basements

Where there is a basement or lower-ground floor, these areas should be provided with alternative escape routes by means of independent protected escape stairways, or directly to the outside in the case of lower-ground storeys. In a building with a single protected escape stairway serving the upper storeys, this stairway should not extend down to any basement or lower-ground storey. The stairway serving the basement or lower-floor storey should be separated from and independent of the stairway serving the upper storeys.

Inner rooms

An inner room is a room where escape is possible only by passing through another access room. Inner rooms are not acceptable for guest accommodation, except in the case of an ensuite bathroom or a small storage room.

Emergency lighting

In the event of a fire, it is possible that the mains lighting system may fail and this would make evacuation of the premises difficult, if not impossible. To provide for this, it is normal to provide a system of emergency lighting to illuminate the escape routes on failure of the mains electrical supply. Emergency lighting should be provided in accordance with the recommendations of I.S. 3217 : 1989 : Code of Practice for Emergency Lighting. Emergency lighting fittings are required in protected stairways and protected corridors/lobbies to indicate the direction of escape. The emergency lighting system should be regularly inspected, tested and maintained, as recommended in the Irish Standard and a record kept in the fire safety register for the premises.

Fire Alarm Systems

A fire alarm system is required in every premises used for guest accommodation, to provide early warning of an outbreak of fire and to ensure that the escape routes can be safely used. A fire alarm is comprised of a number of components, such as detectors, bells, wiring and control mechanism. Some fire alarm systems, such as those based on self-contained units have some of these components combined. Other systems, are more comprehensive, especially in respect of a central control facility. The extent and type of system required depend on the size and complexity of the premises.

Guidance on the provision of a fire alarm system based on self-contained units is contained in BS 5839 : Part 6 : 1995 : Code of practice for the design and installation of fire detection and alarm systems in dwellings. This British Standard refers to different grades of alarms and types of fire detection and alarm systems, providing varying levels of protection. The grade of system (grades A to F) principally concerns the nature of the detector and alarm units employed while the types of the system (types LD1 to LD3) relate to the extent of the detector coverage to be provided. A fire alarm system based on self-contained units is acceptable only in the following situations:

— In a single-storey building with not more than six bedrooms; or

— In a two-storey building with not more than four bedrooms on the first floor.

The grade of system used in guest accommodation should be at least Grade C. This grade is based on the provision of interconnected mains-powered smoke alarms (smoke detector and alarm sounder in a self-contained unit), each provided with an integral standby power supply and provided with an element of central control. The central control facility, which should be located in the entrance hallway, should provide a facility for testing the alarm system, a means for sounding all alarm units simultaneously, and means for the location of an alarm source. The alarm system should be type LD1 as defined in the British Standard. The system should incorporate suitably located and interconnected mains-operated alarms (with integral battery backup) in all circulation areas that form part of the escape routes and in all habitable rooms, including bedrooms, living rooms, store rooms and kitchen. To prevent nuisance activation, a kitchen should only be fitted with a heat detector. It should be noted that Grade F alarms, which are battery-powered only, are not adequate for this application.

Other fire alarm systems

A fire alarm system in accordance with the recommendations of I.S. 3218 : 1989 : Code of Practice for Fire Detection and Alarm Systems for Buildings - System Design, Installation and Servicing should be provided in any building where a system based on self-contained units is not acceptable. Fire alarm systems based on this Irish Standard incorporate manual call points for raising the alarm, alarm sounders, automatic detection of smoke/heat by suitably located detectors, control and indicating equipment, and cabling. The type of fire alarm system should meet the requirements for a type L2 life safety system, as defined in the Irish Standard. An L2 system requires automatic fire detection in all escape routes and in all rooms (with the exception of toilets/bathrooms), including bedrooms and other high-risk rooms located off the escape routes. Smoke detectors should be provided in escape routes while smoke or heat detectors may be used, as appropriate in other rooms. If the guest accommodation is ancillary to some other use, such as a shop or public house, the fire alarm system should cover all parts of the premises (i.e. a type L1 system). The alarm system should be designed and installed in accordance with the recommendations of the Irish standard. The system should be regularly inspected, tested and maintained, as recommended in the Irish Standard and a record kept in the fire safety register for the premises.

Fire separation

The principal method of restricting the spread of fire between different parts of a building is for floors, walls and doors to have an appropriate level of fire resistance. These include the upper floors of the building and walls separating any high fire-risk areas such as laundries, kitchens, storage rooms, etc. The minimum period of fire resistance for such elements of construction is 30 minutes. In a two-storey building with not more than four bedrooms on the first floor, a "modified" 30 minutes (15 minutes integrity and insulation) rating for an existing floor construction would be acceptable. In a building with more than four storeys, the floor construction and other elements of structure should have a fire resistance of not less than 60 minutes. In a building with a basement or lower-ground storey, the period of fire resistance for the ground floor should be 60 minutes. Where guest accommodation is ancillary to some other use, such as a shop or public house, the floor and walls separating the guest accommodation should have a fire resistance of not less than 60 minutes.

Fire resisting construction is also required to protect the means of escape. This generally relates to the walls, including any doors, enclosing the escape routes. In some situations, and subject to certain limits, glazing may be incorporated into fire resisting walls and doors. Guidance on the use of glazing in such situations, and on the fire resistance performance of other elements of construction is contained in Technical Guidance Document B (Fire Safety) to the Building Regulations, 1997.

The fire and thermal properties of the internal walls and ceiling linings of a building can influence the spread and severity of a fire. For this reason, controls are required on the combustibility and surface spread of flame characteristics of all lining materials. Wall and ceiling linings in protected escape routes should have a Class 0 as defined in Technical Guidance Document B (Fire Safety) to the Building Regulations, 1997, while all other areas should have a rating of not less than Class 1 to BS: 476: Part 7: 1987. Any non-combustible material, such as plaster, brickwork, blockwork or plasterboard has a Class 0 performance rating. Combustible materials such as timber will not meet either Class 0 or Class 1 performance. The fire risks associated with combustible wall linings require careful assessment and may require removal or treatment where appropriate to achieve a satisfactory level of performance.

Building Services

Building services such as electrical, gas and heating are potential sources of fire, and equipment associated with them should be installed and maintained in accordance with the relevant standards and codes of practice.

Fire can be caused by defective or inadequate electrical installations or by the use of defective electrical equipment. The electrical installation comprising wiring, sockets, switches, distribution boards and other equipment should be in accordance with the National Rules for Electrical Installations published by the Electro-Technical Council of Ireland. Electrical appliances should conform to a standard appropriate at the time of manufacture such as the relevant parts of I.S. 205: Part 1 : 1980 : Safety of Household and Similar Electrical Appliances (General Requirements). Commercial electrical catering equipment should comply with the relevant parts of BS 5784 : Safety of electrical commercial catering equipment. Emergency lighting should be inspected, tested and maintained in accordance with the recommendations of I.S. 3217 : 1989 : Code of Practice for Emergency Lighting. The fire alarm system should be inspected, tested and maintained in accordance with the recommendations of the relevant standard.

All gas installations, storage tanks, pipe lines, gas burning flues and other equipment should be installed, fitted and maintained in accordance with the appropriate standards, including I.S. 820: 2000 Non-Domestic Gas Installations and I.S. 813 : 1996 : Domestic Gas Installations, as appropriate. Liquefied Petroleum Gas (LPG) cylinders should not be utilised or stored inside any building used for guest accommodation. Gas supply pipes or meter installations should not be accommodated with protected escape routes.

Any tank installation for the storage of LPG should conform to I.S. 3216 : 1988 Code of Practice for the Bulk Storage of Liquefied petroleum gas. The installation should be operated and maintained in accordance with the recommendations of the Irish Standard.

Heating

Heating appliances and open fires are a potential source of ignition. Open fires, where provided in sitting rooms, should be carefully supervised. Bedrooms should not be provided with open fires, or any type of portable fires/heaters. Individual heating appliances, where provided, should be fixed in position. They should not have an exposed flame or heating-element, which could lead to accidental ignition of combustible material. All heating appliances and installations should be in safe working order and be properly maintained. Guidance on the installation of heat-producing appliances, including provisions for adequate air supplies, flues, protection of buildings, etc., is contained in Technical Guidance Document J (Heat Producing Appliances) to the Building Regulations, 1997.

Furnishings and fittings should be of a standard that they cannot be ignited easily or do not contribute to the rapid spread of fire. Items that need to be considered in this regard include upholstered seating, curtains, drapes and blinds and floor coverings.Upholstered furniture should be in good condition and should meet the minimum standard of resistance against accidental ignition for domestic upholstered furniture. Replacement furniture should pass the smouldering cigarette test and ignition source Grade No. 5 as defined in Irish Standard I.S. 254 : 1983 Flame Resistance Requirements for Upholstery. Beds and bed-linen may be ignited by flaming and/or smouldering ignition sources such as matches, cigarettes, radiant heaters or by faulty electric blankets. As a general functional requirement, bedding materials should not easily be ignited, or if ignition does occur, the resultant fire does not spread.

Access and Facilities for the Fire Service

In the event of an outbreak of fire and the fire service is summoned, it is important that fire appliances are able to reach the premises, without undue restriction such as a narrow entrance roadway or gateway that might prevent or delay an effective response. When fire appliances arrive, a suitable water supply is essential to bring a fire under control. Due to the variety of circumstances that exist, it is recommended that the local Fire Service be consulted with a view towards familiarising themselves as to any possible access difficulties and as to suitable sources of water for fire fighting in the immediate vicinity.

Water and Food Safety Management

There are many ways to reduce water usage in the kitchen. Traditionally, saving water has not been a major consideration of commercial food preparers. Many establishments cite the lack of money or employees as reasons for not using water conservation methods. Case histories have shown that water efficiency programs are cost-effective, and most initial costs are retrieved within a two-year period. Participation in water efficiency programs shows that there is a concern for efficient water. These programs are supported by local communities.

Inefficient uses of water in kitchen operations come mainly from two areas: equipment design and behavioral patterns. The main types of water-using equipment found in kitchens are dishwashers, faucets, icemaking machines, and garbage disposal use. Improved technology has eliminated many of the water issues associated with equipment, as more rigid standards have been created to curtail excessive water use.

Water audits of commercial facilities have shown that 60 percent of identified water savings comes from simply installing faucet aerators in all kitchen sink outlets. An effective part of water savings in kitchens is attributed to behavioral patterns in facilities. Awareness programs, education, training, and job performance measures can influence proper behavioral patterns of staff.

All dishwashing machines employ wash, rinse, and sanitizing cycles. The sanitizing cycle typically is the chemical reduction of microorganisms to safe levels on any food utensil. The time taken for a dishwasher to complete a cycle is a combination of mechanical action, water temperature, and chemical action. Most dishwashers use between 2.0 and 7.0 gpm for a complete cycle of cleaning and sanitation. Hot water use varies with the pressure of supply lines, operation speed of the machine, and dish table layout. All these variables are intrinsically linked and any adjustments affect each component. For example, rapid washing cycles necessitate stronger mechanical action and more concentrated detergents for cleaning.

There are four main types of dishwashing machines: undercounter, door, conveyor, and flight. A wide array of models and accessories are available for each category. Requirements for machine size can be calculated by estimating the amount of traffic that will be served in the food service area. Energy guidelines and water consumption levels for dishwashers are becoming stricter, and many manufacturers offer new water-saving models.

Smallest of commercial dishwashers, undercounter dishwashers are best suited for small establishments of about 60 people. They commonly are used in nursing homes, churches, small food service areas, office buildings, and for glass washing in taverns and bars. The undercounter machines are similar to residential dishwashers in that the door opens downward with rack(s) rolling out onto the lowered door for access. A revolving wash arm handles the wash and rinse cycles, with a small holding tank being automatically drained after each cycle. An automatic timer controls cycle length.

Undercounter machines come in both hot water and chemical sanitizing models, with optional booster heaters for the latter. Undercounter machines use the most water per rack of all commercial dishwashers. This illustrates the need to wash only full racks when the machine is in use. Manufactured to service 50-200 people, door type machines are the most widely used of commercial dishwashing machines. Door machines are used in schools, hospitals, churches, restaurants, catering businesses, fast-food establishments, and as glass and utensil units in larger operations. These box-shaped machines have singular or multiple doors that slide vertically for loading and unloading.

Door type machines are available in high temperature and chemical sanitizing models. These "dump and rinse" machines have a single tank for water and detergent, which are circulated in measured volumes and temperatures. Two revolving spray arms (one above and one below the dish rack) distribute wash solutions evenly over the dishes. Some door-type machines now have the ability to recycle rinse water to be reused in a wash cycle.

C-line, or rack conveyor, machines use a motor-driven conveyor belt to move the rack-loaded dishes through a large tank with separate wash and rinse compartments. Most widely-used in hotels, large restaurants, hospitals, schools, and universities, these machines are well suited for service of 200 or more people and accommodating most heavy food service operations. C-line machines come in varying sizes, with available additions such as pre-wash units, side-loading trays, condensers, and blower-dryers. A single tank holds the water and detergent at a regulated temperature.

The wash solution is pumped through multiple spray arms (revolving or stationary) that run constantly once the machine is operational, regardless the presence of a dish rack. The rack is then sent through the rinse compartment, where it is sprayed with the 180 F water by spray nozzles above and below the rack. C-line machines with multiple tanks differ in that some use stationary versus rotating spray arms. The racks then are sent into a pump-driven rinse tank that rinses the dishes heavily. This process usually uses recycled water from the final rinse. All rack conveyor machines have a timer control for the speed of the conveyor to assure proper wash and rinse times.

Water efficient measures, such as the installation of an electric eye sensor (that keeps the conveyor from running when there are no dishes on the racks), have started to make rack conveyors more energy- and costeffective. Some efficient conveyors can reduce final rinse consumption from 300 gph to 130 gph. The use of energy-efficient boosters and low-flow pumps can reduce energy and water consumption levels by 50 percent. Similar in that they use a conveyor belt to move dishware, flight type machines do not have racks. Rather, dishes are loaded directly onto the belt.

Flight type dishwashers provide high-volume washing capability needed only in the largest institutional, commercial, and industrial facilities. Variations in possible machine additions include power scrapers, power wash, power rinse, final rinse, and blower-dryers. Water efficient strategies for these machines include the recirculation of final rinse water, electric eye sensors, extra-wide conveyors, and low-energy built-in booster heaters. These additions have amounted to water savings as much as

47 percent, while maintaining loads of more than 14,000 dishes per hour.

The volume of consumption in dishwashers can be reduced by a variety of practices, all of which target awareness of equipment and operation needs.

Behavioral Modifications

— Educate staff about the benefits of water efficiency and the importance of hand scraping before loading a dishwasher.

— Instruct staff to quickly report leaks and troubleshoot.

— Only run rack machines if they are full.

— Try to fill each rack to maximum capacity.

Mechanical Modifications

— Recycle final rinse water for washing.

— Keep flow rates as close as possible to manufacturer specifications.

— Install "electric eye sensors" to allow water flow only when dishes are present.

— Install door switches for convenient on/off access.

— Check voltage of booster heater to make sure it fits the machine.

— Use " steam doors" to prevent water loss due to evaporation.

— Install low-temperature machines that rely on chemical sanitizing over high water temperature.

— Check volume of service and estimate facility needs ¾ a better option may be a larger machine that has a lower water flow rate per rack.

Kitchen Faucets and Pre-rinse Sprayers

Faucets can waste large amounts of water, as they are the most heavily used water source in kitchens. Conventional faucets, with typical flow rates of 2.5 to 4.0 gpm, can waste as much as 40 gallons of water a day when not fully closed. Since 1994, water efficiency standards have been federally mandated, requiring that all post-1994 manufactured faucets consume a maximum of 2.5 gpm at 80psi. But many facilities have older fixtures with rubber gaskets that wear and deform because of high amounts of hot water use. By simply installing a brass gasket and an automatic shutoff nozzle, a facility could save as much as 21,000 gallons of water per year.

Many adjustments and technology advancements have been made in faucet design as a variety of low-flow faucet types are manufactured. Foot-activated kitchen faucets will reduce water use while providing additional convenience. Faucets used in kitchens will be primarily the conventional type or pre-rinse pressure sprayers. There are a variety of modifications that can be employed for all types.

Water efficiency options for kitchen faucets:

— Adjust flow valve to reduce water flow.

— Check for leaks and worn gaskets.

— Install a flow restrictor to limit maximum flow rate to 2.5 gpm or less.

— Install a 2.5 gpm faucet aerator, maximizing flow efficiency by increasing airflow to the stream.

— Consider infrared or ultrasonic sensors that activate water flow only in the presence of hands or some other object.

— Install pedal operated faucet controllers to ensure valves are closed when not in use.

— Educate staff to look for leaks and broken faucets in their area. n Do not leave faucets on to thaw vegetables and other frozen foods.

— Post water conservation literature and reminders to staff around work areas.

Pre-rinse sprayers are used for rinsing cooking utensils, pots, pans, soaking dishes, and cleaning. They are designed with automatic shut-off valves at the hose head to supply water only when needed. There are water-efficient spray valves offered that supply from 1.6-2.65 gpm at 80psi. These types of sprayers are designed to meet the demands of food service operations.

Ice Making Machines

Ice machines have many commercial uses, from restaurants to lodges, and can use significant amounts of water depending on the type of machine and the desired type of ice. Ice machines are composed of the following components: a condensing unit used for cooling, an evaporator surface for ice formation, an ice harvester, an ice storage container, and, in some models, a dispenser. The type of condenser an ice machine uses will have the largest effect on water use. Two types of condensers are available: air-cooled and water-cooled. Water-cooled machines use 10 times as much water as air-cooled machines and water rarely is recirculated. In comparing water- and air-cooled compressors, the compressor horsepower at design conditions is invariably higher with air-cooled machining. However, operating costs frequently compare favorably during a full year. The desired quality and visual clarity of ice also will influence water consumption. Ice quality, machine cleaning, and water efficiency all need to be balanced for optimum operation.

Garbage Disposals

Studies shows that garbage disposals can waste a significant amount of water. It is recommended that their use be minimized or eliminated from kitchen operations. Many facilities use strainers or traps that employ a mesh screen to collect food waste for proper waste treatment. Another option is to install strainers in sinks, leaving the food matter in the sink for disposal in trash receptacles or composting units.

5

Meeting the Challenges

The hotel and tourism industry has been suffering particularly from the combined effects of a general economic downturn that began to show itself early in 2001 and the shock wave from the events on September 11th of the same year. The general economic downturn had already started to reduce the industry's previously strong 4½ per cent annual growth rate to well below 4 per cent. The impact of the September 11th events on tourism was dramatic.

The whole industry lost an additional 10 per cent or more of its business worldwide, with certain countries experiencing falls of up to 30 per cent. The extreme market lows remained for at least three months, plunging the industry's growth rate for the whole year 2001 into negative territory of somewhere between -1 per cent and - 5 per cent. Although signs of a recovery have been observed since January 2002, it is proving slow to take hold. More recent events have raised new security concerns worldwide. They include the terrorist attacks in Bali, 12 October, and in Mombasa, 14 November 2002.

The Moscow hostage drama, late October, also contributes to an increased feeling of insecurity in the international sphere as does the perspective of a possibly aggravating conflict in the

Middle East. Forecasters had predicted that tourism would return to its pre-September 11th levels in 2003. Recent reports, however, of the renewed decline in the third quarter of 2002 of certain performance indicators, suggest that a longer recovery period is to be expected. A flat performance, i. e. continued stagnation, now seems more likely. Longer-term predictions now suggest that the industry's pre-2001 growth rates will not be regained before the year 2004 and that it will take the market until 2005 to get back to its - outstanding - year 2000 volume.

Employment figures in the sector have reacted more slowly to the crisis, but the negative effects will be longer lasting. Employers at first hesitated to reduce staff levels, although reducing costs on personnel through shorter hours and lower fringe benefits. The common tendency, particularly in the service sectors, is for staff lay-offs to be partially compensated by increased productivity. Consequently, the longer the crisis lasts, or the slower the industry recovers, the more jobs are lost irretrievably. According to estimates by the World Travel and Tourism Council (WTTC), the events of September 11th and subsequent developments will have caused more than six and a half million jobs to be lost in the travel and tourism industry over the years 2001 and 2002 - some 6.9% of the total employed, as compared to a normal growth that could have been expected in normal times without the impact of a crisis. This estimate does not include those related sectors providing inputs to travel and tourism, where employment may have suffered in similar proportions.

Although international tourism is generally highlighted for statistical convenience, the crisis has had an important impact on domestic tourism levels as well, particularly in the United States, where much of the internal tourism involves the use of air transport. Both international and domestic tourism tend to reflect the general level consumer confidence. Travel and tourism are the types of product on which consumers can easily make substantial savings if they need to, e.g. by purchasing lower quality services, by taking shorter leave or by spending their vacations closer to home. Adjustments like this have undoubtedly been made by consumers in order to adapt to the present increased economic and psychological pressures on travelling.

Nevertheless, there have been some initial signs of recovery in leisure tourism as can be observed in tour operators' business in early 2002. These signs serve, in a way, to illustrate both the extent to which leisure tourism suffered immediately following September 11th and the fact that the crisis phase immediately thereafter in this particular sector seemed actually to be quite short. One contributing factor might have been that the promotional campaigns undertaken by the tourism industry to mitigate the negative effects of the crisis, accompanied by measures taken to support small enterprises. They have also had a much greater effect on leisure tourism than on business travel. However, general economic constraints continue to force consumers to apply various types of savings measures such as avoiding air transport, choosing cheaper destinations, staying for shorter periods of time, and booking late.

Another point worth noting is that developing countries are not benefiting proportionately from any recovery in leisure tourism because of their heavy dependence on foreign tourists, and particularly on long-haul tourists. Business travel (which falls under the generally accepted definition of tourism if one night is spent outside the usual living sphere) is recovering more slowly, if at all. Corporate travel professionals are not expecting to see a recovery in business travel for at least another year. Although in terms of numbers, business travellers may account for no more than one-fifth of the total travelling population, they contribute nearly one-half of the industry's total revenue because of their high frequency of travel. It has been observed that, in light of the economic downturn, travel managers of corporations and businesses are moving to re-negotiate contracts in order to reduce business travel costs. Most affected countries have taken measures on a central, regional or local level, aimed at mitigating the effects of the tourism crisis. To varying degrees, losses in long-haul international tourism have been successfully compensated by gains in international tourism from within a region. This is the case with certain European countries and with Canada, and some therefore ended the year 2001 registering positive growth in their levels of international tourism.

Domestic tourism in many developed countries benefited from reduced outbound tourism because many clients who were previously long-haul tourists now favour surface travel, and cheaper, shorter trips. Major and, at times, remarkable measures were taken to stimulate consumer demand and support the industry by sectoral bodies, including national tourist boards and ad hoc committees composed of tourism authorities and industry representatives in public-private partnerships. Governments have been participating financially in advertising campaigns, quality improvement programmes, tax relief schemes, or the granting of subsidies to small and medium-sized enterprises (SME's). Despite the fact that the developments in tourism after 11 September 2002 are those of a crisis with a beginning and hopefully with an end, they seem to have introduced or at least accelerated lasting changes in consumer demand for tourism products. Specialists agree that patterns such as "sea, sand and sun" may get out of fashion as may the desire of many tourists to travel as far as possible and to places as exotic as possible. New trends show an inclination to stay closer to home and make sure that holidays are an "experience," be it in the area of cultural enrichment, in the area of sport or other. Providers are challenged to meet those specific types of demand by developing higher value products and at the same time compensate a tendency of consumers to spend less money. Developing countries are particularly challenged in order to compensate for long distance travel, language differences, and less complete infrastructure to receive tourists other than in beach resorts.

For most countries, knowledge of the volume of border crossing tourism is largely limited to the numbers of arrivals and the expenditure by international tourists in the host country published with varying delay by governments and presented annually by the World Tourism Organization (WTO). Some more frequent statistical surveys are carried out based on samples covering also domestic tourism and especially hotel performance indicators, but these do not produce figures representative of the total size of the industry. Only a dozen countries have so far started to develop statistical instruments to measure the whole tourist industry (Tourism Satellite Accounts), and at present results are published only after long delays. Estimates that are available are based on different methodologies and hardly comparable. When it comes to employment figures, however, there are hardly any other sources available. Another complication is that definitions of the 'tourism' market or industry vary according to the particular perspectives of the different organizations. For example, is 'travel' included or not? (It is by the WTTC).

Direct Impact of September 11

In 2001 until September 11th, travel and tourism had been growing at the annual rate of 3.8 per cent and was expected to maintain that level of growth for the full year. In the event, according to WTTC estimates, travel and tourism demand worldwide in 2001 ended 1.3 per cent down on the previous year. The real decline for the whole of 2001 as compared to the expected growth was therefore 5.1 per cent, of which it has been estimated that as much as 4.4 percentage points can be attributed to September 11th alone (85%). In 2002, the year-on-year decline of travel and tourism was expected to continue, with WTTC forecasting last spring a slight decrease by 0.4 per cent for the whole of 2002. The forecast was adjusted in fall when WTTC estimated a flat performance as compared to 2001. On the assumption that last year's pre-September 11th growth of 3.8 per cent would normally have continued throughout 2002, this would mean another real annual decline for tourism of close to 4 percent, of which around 3 percentage points can be attributed to the events of September 11th alone (70%).

Receipts from cross-border tourism in 2001

The positive growth trend that prevailed throughout the 1990's was brought to an end in 2001 by a drop in the industry's income from international travellers of some 5.2 per cent at constant US Dollar prices. It should be noted that the receipts from international tourism dropped deeper than the number of international arrivals, shown in the next section, due to new general consumer preferences for shorter stays and cheaper services.

Border-crossing tourist arrivals in 2001

In an attempt to separate the effect of September 11th from the effect of the general economic downturn in 2001, the first eight months and the last four months of the year are shown in different columns. The number of international tourist arrivals worldwide fell by 0.6 per cent in the full year 2001, compared to the year 2000, but there were important differences between regions and countries both for the full year and for the four-month period after September 11. The worst losses were felt in the Middle East and the Americas, particularly North America, where international tourist arrivals were down by 6.8 per cent in the whole year 2001, but by as much as 22.6 per cent in the last four months compared to equivalent periods of the previous year.

Countries most affected by the September 11th events were the United States, with international tourist arrivals having fallen to their lowest point between September and November, on average over 30 per cent below the level of the same period the previous year. Countries in the immediate neighbourhood of the US also received significantly less tourists than at the same time the previous year, e.g. Canada: (-19%), Cuba (-26%), Dominican Republic (-25%), Mexico (-24%) and Jamaica (-20%). In general, all countries that would normally have expected a high proportion of their incoming long-haul tourism to originate from the United States experienced steep declines in international tourism in the first three months following September 11.

In Europe, those worthy of note include the United Kingdom (-12 %), Germany (-17%), Switzerland (-16%), Italy (-11%) and Austria (-9%) - and elsewhere, the Philippines (-

25%) and Australia (-21%). Some other countries amongst those whose tourist industries are important elements in their national economies suffered badly such as Egypt (-55%), Tunisia (-25%) and Morocco (-22%). These preliminary figures served to confirm earlier predictions that the three types of tourist destinations likely to be most affected by the events of 2001 were:

— countries highly dependent on US outbound travel;
— destinations a long distance from their main generating markets; and
— countries of the Muslim world

In addition to security aspects, the general economic downturn in 2001 also contributed to the poorer performance of international tourism in 2001 compared to the previous year. Here, the main causes are to be found in those major sending countries such as the US, Germany, and Japan. South American markets suffered from economic instability in Argentina. An additional factor for the US as a receiving country, which lost as much as 2.5 per cent on international arrivals in the first nine months of the year, was its strong currency. Among the countries performing worse as a destination for the full year 2001, compared to the year 2000, were the US (-12.6%), Argentina (-9%), Brazil (-8%), the UK (-6.6%), Mexico (-5%), Germany (-5.5%), Italy (-5%), the Dominican Republic (-5%), Jamaica and the Bahamas (both -4%). Other countries experienced a significant drop in their 2001 total tourist numbers year-on-year because, rightly or wrongly, they were associated with security risks, and this was for various reasons not necessarily connected to the September 11th events. These countries included Egypt (-16%), Nepal (-22%), and Sri Lanka (-16%).

By contrast, there were some countries, notably in Europe bordering at the East Mediterranean Sea and in Southern Europe that in 2001 continued to receive higher numbers of foreign tourists than in the previous year, probably because they provide a convenient alternative to long-haul destinations for many European tourists. Thus Turkey (+12%), Croatia (+12%), Slovenia (+11%), Spain (+3%), Greece (+2%) and Cyprus (+1%) all benefited in 2001. It was also observed that several emerging tourist destinations in Europe showed strong growth in the full year 2001, notably Bulgaria (+14%), Slovakia (+13%) and Estonia (+9%). France (+1.2%) also recorded some growth in 2001 versus their previous year totals, benefiting from their familiarity and proximity with European holidaymakers and even compensating a temporary 20 per cent drop immediately following September 11th which was probably due the sudden lack of Americans visiting Europe.

Australia and New Zealand reported a short period of decline in inbound tourism immediately after September 11th, but most of these losses were partly offset by something of a boom in numbers of foreign tourists later in the year. Some countries, such as the UK, Germany, China and Canada continued to send record numbers of tourists to this part of the world throughout 2001. They did not prevent, however, a year-on-year decline of 2.7 per cent in numbers of international arrivals, mainly accounted for by 4th quarter decreases from Japan (-28.9%) and the US (-17.5%). Notably, the region with the best performance as tourist destination in 2001 was South East Asia and the Pacific, which in spite of a decline after 11 September still ended the year with a growth of 5.5 per cent, mainly due to intra-regional tourism. Performing regions were also Africa and Central America, albeit on a small basis in absolute terms.

The United States

The United States is the country whose hotel, travel and tourism industry was most seriously affected. According to the US Office of Travel and Tourism Industries, the total number of international arrivals fell by almost 11 per cent over the whole year 2001. This was largely because of a dramatic 30 per cent drop in September, moving to a 34.0 per cent loss in October, with an average 31.8 per cent loss of business over the last four months of the year compared to the same period of the previous year. October was the worst month. The numbers of international arrivals slowly started to recover in November (Table 1).

Table 1

Non-resident arrivals to the U.S. By world region or country of residence 2001

	Total	*Canada*	*Total overseas*
Change from same period the year before (per cent)			
October	-34.0	-24.3	-39.3
November	-29.2	-21.0	-34.1
4th Quarter	-31.8	-22.8	-36.9
Year 2001	-10.7	-7.8	-13.67
Count (million)			
Year 2001	45,5	13,5	22,4

Certain countries supplying travellers to the US performed worse than others because they were more seriously affected by economic crises. These included Brazil (-49%); Germany (-46%) and Japan (-45%). But all of the top 10 markets experienced steep falls of around 20 per cent or more, as is shown in Table 2 below. By the end of the year, however, it was the German market that had performed worse than any other, which is probably due to generally nervous consumer behaviour in the light of the country's poor economic performance.

Table 2

Top 10 markets for tourism to the US
Arrivals Change in 2001 and projection for 2002

	2001/2000 (Sept.) (Per cent)	*Year 2001/ 2000 (Per cent)*	*Year 2002/ 2001 (Per cent)*	*Count 2001 (million)*

Canada	-19	-7.8	0.9	13.5
Mexico	-30	-7.4	2.6	9.6
United Kingdom	-27	-10.7	1.5	4.2
Japan	-45	-18.5	-3.6	4.1
Germany	-46	-24.6	-1.6	1.3
France	-34	-14.0	-1.6	0.9
South Korea	-20	-5.3	5.1	0.6
Brazil	-49	-22.7	-1.2	0.6
Venezuela	-21	-1.1	-8.6	0.6
Italy	-40	-19.8	3.7	0.5

Demand for accommodation in the US is estimated to have suffered a far more modest decline than international arrivals, another indication that domestic tourism was less affected and it was long-haul travel that suffered most. Accommodation demand had already declined by 6.0 per cent in the third quarter of 2001, as compared to the same period in 2000. However, the decline in the fourth quarter was markedly greater (6.5%), but still only a fifth of the decline in international arrivals.

Early forecasts by the Travel Industry Association of America already suggested that after the 7.6 per cent drop in 2001, the recovery in consumer expenditure levels on travel in the US would be slow in 2002, with an annual growth rate of only 2.2 per cent. Leisure trips in March, April and May 2002, for example, were 4 per cent below the level in the same period a year ago.

Overall, first quarter 2002 international arrivals to the US were down 12 per cent (8.3 million compared to 9.5 million in 2001). Four of the top markets - the United Kingdom, Germany, the Netherlands and Switzerland - posted declines for the first quarter of 2002, but the declines are less than those recorded in the pre-September 11th months in 2001. Mexico and Spain are the main markets posting positive growth in numbers of arrivals through the first quarter 2002, at +22 per cent and

+6 per cent respectively versus the same period last year. Recovery in the US is not being enjoyed equally by all sub-sectors of the industry.

Leisure tourism services are recovering faster, for example, than the sub-sector catering for business travellers. In both the US and in Canada, where large advertising campaigns featured domestic attractions, resort destinations are being filled by leisure tourists driving their cars instead of flying. However, on another dimension, according to a sample of hotels monitored in the USA, the fullservice hotels suffered a decline in their Revenue per Available Room (RevPAR) of 13.5 per cent in the first six months of 2002, although operating expenses were reduced by 9.1 percent as compared to the same period in 2001. Profits (defined as income before deductions for capital reserves, rent, interest, income taxes, depreciation, and amortization) declined as much as 21 per cent.

The Marriott hotel chain had publicized similar results at an earlier stage. The continued decline in the more profitable business travel had forced its RevPAR down by an average of 12.7 per cent. Marriott said its earnings for the first quarter 2002 were down 32 per cent versus a year earlier, on income that had dropped by 4 per cent. The US accommodation industry is estimated to save US$3.4 billion in 2002 through cost-cutting measures that were put in place in 2001. Without these actions, profits would have decreased by 18.5 per cent to US$13.7 billion in 2002. However, more recent developments seem to limit recovery expectations for 2002. The negative wealth effect of the recent sharp decline in stock values, the general erosion in consumer confidence in the US, increasing frustration with the inconvenience of air travel and heightened fears of a war against Iraq, will restrain lodging demand growth and encourage continued discounting through the end of 2002. Marriott reports, for US properties across its lodging brands, a decline in RevPAR by an average of as high as 6.8 percent in the 2002 third-quarter. Price water house Coopers forecast RevPAR in the US to decline by 2.3 percent this year but to increase by 3.5 percent in 2003. The more robust outlook for lodging has, however, been delayed to 2004. In particular the business travel industry is forecast to continue its downward move this year with income falling 4.3 per cent. It is expected to stabilize in 2003 with a gain of still less than 1 per cent. Overall travel expenditures in the US are expected to virtually stagnate in 2002 and strengthen to an annual growth rate of 5.0 per cent by 2003 (Table 3 below), reaching US$555.6 billion, still below that of the year 2000.

Table 3

Travel Expenditures in the United States

	2000	*2001*	*2002*	*2003*
Total (US$ billion)	570.5	537.2	535.3	555.6
Annual Growth (per cent)	7.1	- 5.8	-0.4	5.0

The Travel Industry Association of America observed that by August 2002 the annual rate of decline in US hotel and motel receipts was running at - 1.3 per cent. However, the trend was positive, because the decline was as high as - 5.9 per cent for the month of June 2002. The decline in the level of employment in hotels and motels, however, improved at a slower pace, with as high as - 4.7 per cent in August against - 5.2 per cent in June. Employment in eating and drinking places fell by -1.8 per cent in August, even worse than in than in June (-1.3) and contrasting with the positive, albeit equally declining, levels of growth in receipts of 3.7 per cent and 5.5 per cent respectively (see Table 4).

The drop in outbound tourism from the USA is still being felt in many countries, but most strongly in Europe. The number of American tourists in France in April 2002 was almost one-third less than in the same month a year earlier. In the United Kingdom, the number of American tourists in the early months of 2002 was 10 per cent less than in 2000 - the last "normal" year before a combination of foot-and-mouth disease and fears of terrorism sent American and most other visitor numbers into free fall. And the number of American visitors was down 11 per cent in the first half of this year in Rome, Florence, Venice and Naples, compared to the same period a year before.

Table 4

Hotels and Restaurants in the USA

Annual growth rates for June and August 2002/2001

	Hotel/Motel		*Eating and drinking places*	
	Receipts	*Employment*	*Receipts*	*Employment*
June 2002/June	- 5.9	- 5.2	+ 5.5	- 1.3
August 2002/August 2001 (per cent) 2001 (per cent)	1.3	- 4.7	+ 3.7	- 1.8

Japan published figures recently, as shown in Table 5, based on reports by Japanese travel agents indicating that by November 2001 they had suffered a drop of 21 per cent in their total business since September 11th. They reported a record level of cancellations representing 52.1 per cent of all overseas package tours with departure dates between September and November.

The number of Japanese travelling abroad was down 9.1 per cent at the end of 2001 versus the previous year, according to a report released in January 2002. Whereas in May 2002 the number of bookings of foreign trips through Japan's top 50 travel agencies was still 6 per cent less than the previous year, July and August, the peak season, have been especially slow, with 10 to 20 per cent fewer bookings than in the previous year. By contrast, the numbers of visitors to domestic destinations were up by 4.6 per cent to 2.07 million. In the period immediately following September 11th, the largest travel agent (JTB) saw an increase of 19 per cent in their bookings for domestic tours departing during the October-December period.

Table 5

Travellers within, to, and from Japan

	2001 estimate		*2002 prediction*	
	Thousands	*Growth (per cent)*	*Thousands*	*Growth (per cent)*
Total	341,310	-0.6	340,130	-0.3
Domestic Travellers	325,110	-0.1	323,490	-0.5
Overseas Visitors to Japan	4,730	-0.6	5,140	8.7
Travellers to Overseas Destinations	16,200	-9.1	16,640	2.7

China experienced some impressive growth in both domestic and inbound tourism, with domestic tourism developing particularly strongly. Total annual revenue from tourism has grown on average by 12.7 per cent over recent years, much faster than the gross domestic product (GDP) which grew by an average rate of 7.4 per cent. More than 73.34 million overseas tourists visited China during the first 10 months of 2001, an increase of 5.7 per cent on the same period of 2000, or almost double the world's average growth rate for 2001 before September 11th.

The total amount spent by overseas tourists in that period grew by 9.3 per cent to US$14.8 billion. Over two-thirds of the revenue is still, however, from domestic tourism. Spain's tourist industry, which directly employs more than a million people and accounts for 12 per cent of the economy, is suffering a protracted crisis, recording 10 per cent less visitors for the first six months of 2002 compared to the same period the previous year. The number of German visitors to Spain in July 2002 was 20 per cent lower, pushing down the hotel occupancy rate by 5.5 per cent. Reservations made by German tour operators tumbled by as much as 37 per cent. The crisis is attributed partly to growing unemployment in Germany, but also to price increases in Spain, with German tourists preferring cheaper destinations such as Turkey and Croatia.

In Switzerland, tourist numbers for the six-month winter season 2001-02 were a dramatic 6 per cent down on the previous year's winter season result. Yet this was still 1.4 per cent above the average for the previous five winter seasons. Increased domestic tourism seems to have stepped in to replace international visitors, whose numbers dropped by 10 per cent. The number of tourists visiting from the US was nearly 20 per cent less and from Japan 31 per cent less than in the same period the year before.

Overnight trips to Canada were up by 0.6 per cent for the first five months of 2002 compared to the same period in 2001, boosted by an increase in US travellers entering Canada by car (accounting for 9.9% of the total). On the other hand, the number of Canadian tourists to the US or overseas was still down by 7.2 per cent (at 1.5 million) in May 2002, which represented something of a recovery because it was down 11.3 per cent for the five months before, compared to the same period a year ago. The government of Morocco predicts that tourism revenues will have dropped by US$700 million by the end of 2002, down by one-quarter on the total revenue of US$2.7 million reached in 2001.

Morocco's tourist industry recorded a 43 per cent revenue drop in January 2002 compared to January 2001. The industry currently employs over half a million people, serving some 2.4 million foreign tourists annually. Major markets supplying most of Australia's tourists showed dramatic falls in outbound tourism during the 12 months to July 2002, with the number of visitors from neighbouring New Zealand dropping by 25.5 per cent, from the UK by 17.7 per cent and the United States by 8.4 per cent, according to the Australian Bureau of Statistics. Overall, international visitor arrivals have fallen 8.4 per cent to 409,400 from January to July 2002, compared to the 446,900 tourists for the same period last year.

These results were attributed September 11th mainly, but Australian tourism leaders feel under pressure by the lasting downturn, leading them to reconsider their whole marketing strategy. A return to pre-September 11th levels of international arrivals was still seen as being a long way off. After the bomb attack of October 2002, the Australian tourism industry fears negative effects of tourists avoiding Asia. Singapore and Bangkok are stopovers on the most popular route from Europe to Australia. Any disruption to the flow of tourists to them would have a severe knock-on effect on Australia. Industry reports in late October 2002, however, suggested the tourist numbers from all markets would remain largely unaffected as a result of the 12 October Bali attacks.

EMPLOYMENT

Employment numbers in the tourism sector do not react immediately to a decrease in business because the special characteristics of employment relationships in the sector allow for some labour cost reduction to take place before workers finally become unemployed. Flexible working arrangements such as part-time or on-call working are common for at least one quarter of the labour force and help to avoid their complete unemployment in a business crisis, albeit with reduced income. Employees accounting for almost another quarter of the total have full time but temporary employment contracts and are the first victims of formal unemployment when reducing their working time and income becomes insufficient to address the income crisis of the enterprise.

Employers usually hesitate, however, to reduce their core of trained staff. Only if a crisis persists, businesses tend to re-structure and workers are eventually laid off permanently. Jobs then remain lost, even in times of recovery. As described earlier, the WTTC estimates a 7.4 per cent drop in tourism demand over the years 2001 and 2002 combined, instead of the 3.8 per cent growth that had been anticipated previously for each of these two years. The sector is now seen as being in a state of stagnation, which is a dramatic change in fortune considering its previously very high growth rate.

The WTTC estimated that over 6 million jobs would have been lost in the tourism industry worldwide in the years 2001 and 2002, which represent 8.6 per cent of employees, or one in every 12 jobs in the industry. The process of attempting to derive employment estimates from demand estimates has to take into account that normal growth may not result in increased employment because normal growth is achieved largely through gains in labour productivity. WTTC estimated therefore that a growth of demand by 3.8 per cent would have entailed a growth of employment by only 2 per cent. Given this proportion, the dramatic effects on employment forecasted by WTTC can be better appreciated. WTTC foresees a modest recovery of employment in the travel and tourism industry worldwide in 2003. However, with a growth of 1.7, the year 2003 will still end with a 6.4 million jobs lost as a result of the crisis compared to the expected normal annual growth rate of 2 per cent in that year.

Table 6

Net Job Losses in 315 Metropolitan Areas of the United States (Thousands)

	2001	*2002**	*2003**	*2004**
Eating & Drinking Places	-17.95	-150.56	-174.18	-105.22
Hotels & Motels	-49.17	-140.61	-99.40	-38.64
Miscellaneous Amusement & Recreational Services	-33.00	-100.36	-76.28	-42.30
Scheduled Air Transportation	-46.16	-85.89	-35.19	-28.63
Total of core tourism related sectors	-145.83	-477.42	-385.05	-214.79
Total Job Loss (all sectors)	-172.53	-758.92	-587.07	-340.97

The Travel Industry Association of America (TIA) estimates that 19 million Americans were employed directly or indirectly by the travel industry, with an annual payroll of $174 billion. The year 2001 ended with domestic travel volumes and expenditure down 3.5 per cent and 7 per cent respectively, and volumes and expenditure from international travel to the US down 12.6 per cent and 11.2 per cent respectively. Employment in the whole industry in the USA was down 5.8 per cent for the year, with an estimated 1.1 million jobs lost.

High volume tourist destinations such as Las Vegas, Orlando and Honolulu are expected to have suffered major losses in 2002 due to cutbacks in travel. Among 315 metro areas examined by the Milken Institute at the end of 2001, many jobs would be lost as a result of the events after 11 September 2001 in New York, Los Angeles and other big cities.

A ranking of industries (at the 3-digit level of the International Standard Classification of Industries) most affected by job losses after September 11th shows the hotel and catering sector represented by the top two categories listed (see Table 6). The Milken Institute forecasted heavy job losses for the years to come as well, although at a decreasing rate. Over twothirds of the estimated 760,000 total direct jobs expected to be lost in US metropolitan areas were in core sectors related to travel and tourism.

THE YEAR 2002 AND BEYOND

Taking a longer-term view, it may be concluded that the September 11th events merely served to exacerbate an already established downward trend by adding, for a limited time only, some extra ingredients such as the perception of heightened insecurity. The general economic downturn in 2001 had already started to affect all forms of consumption and travel and tourism was probably one of the sectors that suffered most because it is a market in which consumers can easily make savings. Indeed, future variations in the level of consumer expenditure on tourism will probably be driven mainly by variations in levels of present and expected household income.

The Travel Industry Association of America (TIA) undertook three Travel Confidence Surveys in the US, one from 21 to 22 September, another from 12 to 13 October 2001 and the third from 2 to 4 November 2001. In these polls, only 14 per cent of respondents cited safety as their top travel concern. On the other hand, 42 per cent cited economic concerns as reasons for not travelling. Nonetheless, the fear factor continued well into 2002 as a strong deterrent to travel in the US and the health of the industry continued to be affected by federal alerts, air-travel-related accidents and other concerns.

A German survey of 8,000 households showed that in March 2002 only one person in ten claimed to make travel decisions on the basis of security considerations, compared to November 2001 when the proportion had been almost one person in three. It has been predicted that tourism would recover slowly and regain the levels of before September 11th by the third or fourth quarter of 2002. This forecast would be in line with experiences from previous security crises affecting tourism, albeit local, whose effects lasted for about one seasonal cycle.

The continuing economic slowdown in the western world, as well as more recent terrorist activities targeting tourists in Djerba, Bali and Mombasa, however, are having an added effect on the tourism crisis. The year 2002 has been one of rather little growth or even slight decline of tourism performance indicators, with the possible exceptions of the Asian Pacific region and South Africa which has recorded double digit growth of tourism out of Europe in 2002, or China which does continues to grow rapidly both as a destination and as a market. Estimates by large tour operators are usually good indicators of the volume of business in leisure tourism, which represents roughly two-thirds of all travel and tourism activities. Preussag, a major German tour operator, was expecting the poor performance of the winter season 2001-2002 improve for the summer season. By June 2002, however, summer bookings in Germany were down 14.2 per cent on a year earlier, even worse than in winter, when total Preussag bookings were down by 10.9 per cent.

Surprisingly, the German Travel Agents and Tour Operators Federation expected much better results in financial terms than in numbers of bookings, which suggests a scenario of structural changes in the leisure travel business leading towards higher quality demand. According to German tour operators, because of higher price levels, total turnover for the 2000/01 season which ended in October grew by twice as much (+ 6.8%) as the number of bookings (+3.2%), reaching a total of Euro17.4

billion. For 2002, despite bookings being down, German tour operators expected a turnover close to that of 2001, mainly because certain customers are making higher value choices even if they go for shorter stays.

In general, tourism demand was in the last 12 months or so subject to important structural changes including a re-distribution of demand between countries and regions. Leisure destinations located close to important markets have prospered; domestic tourism has made a difference; long-distance international air transport has suffered more than international tourism in general due to increased short distance trips across borders; business tourism and conventions, luxury hotels, and cruises during the winter season were particularly affected, while social tourism, rural tourism, rail and road travel made gains. This is why the decline in business revenues was generally sharper than the decline in the number of travelers. Among the destinations receiving European tourists, short-haul destinations have generally benefited at the expense of long-haul destinations, with demand for North America showing the greatest fall in 2002 so far (-13%). The main longhaul destination to buck the trend has been China, but a number of other Asian destinations and South Africa have similarly recorded double-digit growth out of Europe this year.

International arrivals to countries in the European region will show little change in 2002 over last year which closed with a decrease of -0.6% in international tourist arrivals in Europe. Preliminary figures for the first six to eight months of 2002 indicate a small decrease in the region of -0.2% to -0.3%). As of November 2002, preliminary results show that Northern Europe has suffered the biggest decline in tourist arrivals in Europe. The other sub- regions have been more successful, especially the Mediterranean countries and 'new' destinations such as Turkey, Croatia and Bulgaria.

Austria, as a typical short haul destination for European markets saw foreign leisure arrivals grow by a massive 12% in the first half of 2002. This growth, however, is expected to have eased to a rate of only around 2% for the whole year.

For France, preliminary data on international arrivals (classified hotel accommodation only) suggest that the lucrative US market remained down by no less than a fifth in the first eight months of 2002 as compared to the same period in 2001, whilst growth in the Canadian market was replaced by decline. All international arrivals together performed at a low 0.5 per cent growth, as the performance of the major generating markets within Europe has been considerable with tourists from the UK (almost half of foreign clients of classified hotels) increasing by close to 7 per cent, and arrivals from Germany (another quarter) still increasing by 14 per cent. The nascent trend towards short-haul, short -.break holidays seems to be re-configuring the intra-regional European market, with France being perhaps the main beneficiary.

Germany as a destination lost 3.3% in the number of foreign overnight stays for the first half of 2002, according to preliminary data, which, however, was still 0.2% higher than in the same period in 2000. Most of the loss is due a decrease in long haul tourists such as from the Americas (-14.6 %). Arrivals from within Europe suffered much less and still grew by 0.9 per cent. Those markets whose growth is worthy of mention in the first six months of 2002 are the Netherlands (2.9%), Switzerland (4.0%) and Norway (6.7%), Russia (9.3%) and China (13.1%). Still, Germany remains out of favour with some important markets, such as the UK (-6.7%), France (-3.4%), and Italy (-2.3%). Foreign travelers to Italy remarkably did not decline substantially, but expenditures did. They dropped by as much as 11 per cent in the first six months of 2002 as compared to the same period in 2001. This phenomenon is likely to exist also elsewhere, but figures on 2002 are still scarce.

A slight growth of foreign arrivals to Italy is anticipated for the remainder of 2002 which will close with a growth of 0.9 per cent. Tourist arrivals at Spain's frontiers increased in the first two quarters of 2002 by comfortable 6.9 percent, but only by 0.2 per cent in the second quarter as compared to the same periods in 2001. The summer period may not have been any better. It was, however, in the United Kingdom, where after a weak winter and spring period, July 2002 arrivals exceeded those of the year before by as high as 15 per cent and even those of July 2000. Expenditure increased less, but still by 7 per cent.

Estimates on the whole of 2002 see arrivals increase by comfortable 6 per cent, but expenditure by as high as 9 per cent indicating that off season tourists spend more money than the (younger) summer holiday travelers.

Canada had continued growing as a destination for foreign tourists during the crisis year 2001, albeit at a slower pace, since most of foreign travelers came from the USA and increasingly used surface transport, in part substituting Canada for a long haul destination elsewhere. In the first six months of 2002, however, the number of foreign tourists from the USA slightly shrunk (by 0.2 per cent), and the receiving of overseas tourism, which had been down in 2001 by 8.4 per cent (arrivals) and 4 per cent (expenditures) continued declining by 13 per cent (arrivals), reducing anticipated overall growth of foreign arrivals in 2002 to only 0.1 per cent.

Japan receives one international traveler only for five Japanese outbound tourists. Out of the modest number of less than 5 million that do arrive, it is estimated that over a quarter are in Japan on business and that half of this market originates from Asia. Having suffered in 2001 only a slight 0.3 per cent decrease of the annual growth rate for all tourism, with a higher growth rate for leisure tourists of 0.9 per cent, Japan is estimated to benefit in 2002 from an impressive growth in foreign arrivals of 16 per cent, to a large extent due to an increase of the number of tourists from China.

Outbound tourism from Japan, important for many destinations in the world, is currently forecast to record modest growth in 2002 of a maximum 1-2% after a 9% decline in 2001. The Pacific Asian countries have been doing well as destinations in spite of the world crisis in 2001 and 2002. High GDP performances in mainland China, India and South Korea suggest the trend of Asian intra-regional tourism will continue.

India as a source country appears to record strong double-digit growth in 2002, albeit on a small basis. Inbound tourism in India, however, after dropping in 2001 by a moderated 4.2 per cent (arrivals) and 5.5 percent (expenditures), continues suffering from the conflict with Pakistan and the closeness to Afghanistan. Northeast and Southeast Asia are driving the region's growth - both as destinations and as markets. During 2001, almost 65% of international visitor arrivals in Northeast Asia came from within Northeast & Southeast Asia. For destinations within Southeast Asia the proportion was over 64%.

China has continued to show strong growth in 2002. In the first eight months of 2002 foreign arrivals in mainland China increased by more than 16%. Good increases of inbound tourism have also been registered by Thailand and Vietnam. However, the impact of the Bali tragedy on the region's tourism remains to be seen. Elsewhere in the developing world, only South Africa and Central America and perhaps the Caribbean are expecting slight increases of inbound tourism in 2002 over 2001's level. The USA has been one of the source markets hardest hit by all crisis factors together: the events of 11 September 2001, the continuing threat of terrorism in 2002 and the economic recession. Outbound trip volume fell by 12% in the first half of 2002. US outbound travel demand expressed in terms of total spending is forecast to fall by a relatively modest 2% for the full 12 months of 2002, after a decrease of 5% in 2001. As a destination, the Americas overall are expected to still suffer a 6% drop in demand in 2002.

Some general estimates anticipate for 2003 a slight growth of tourism worldwide of up to 2 per cent. A strong growth of 4 per cent or more is predicted for 2004. Such forecasts seem to have the unanimous support of the specialist community, but the risks are also clear: they depend on the absence of further terrorist attacks; more selective reactions by consumers who should not stop travelling altogether when security threats in fact are limited to a location; the development of the conflict in the Middle East and its oil price implications; and last not least on the general prospects of disposable income. In any case, forecasts by renowned specialists and organizations seem to be highly insecure.

Due to the continuing weakness of the world economy, travel and tourism demand in 2003 is likely to be substantially slower than projected earlier, with business travel in particular hit hard by the pressures on companies to cut costs and bolster their profitability. International visitor spending is now expected to recover by a modest 1.6% in 2003 after falling 10% over the previous two years, leaving international visitor spending in 2003 more than 20% below what might have been expected if it had grown by a more typical 5% a year since 2001. For foreign tourist arrivals to certain European countries in the years to come, forecasts indicate different growth rates. Austria will remain at a modest 2 per cent growth of foreign leisure trip arrivals in 2003.

A modest turnaround, with growth of 2% to 3% is expected for the next 12 months also in Germany. In France, international arrivals are expected for 2004 to be back to the pre-2001 growth rate of over 4 per cent, albeit under the assumption that long haul tourism to France will recover. Foreign visitors to Italy in 2003 are expected to increase by 2.7%, making a total of 39 million. Likewise, a growth rate between 2 and 5 per cent is assumed for the near future in Spain. For the USA, the expectations of the Travel Industry Association of America concerning the growth of international visitor arrivals and expenditures in the years to come are reflected in Table 7. They appear rather optimistic as far as the expenditures are concerned.

Table 7

USA: International Visitor Arrivals and Expenditures

	2000	*2001*	*%*	*2002*	*%*	*2003*	*%*	*2004*	*%*
Arrivals (million)	51.3	45.7	-11	46.7	2	50.6	8	54.4	8
Expenditure (USD million)	82.000	72.300	-12	76.400	6	84.200	10	92.400	10

Forecasts for Canada's foreign tourist arrivals see a growth of 3 per cent each in 2003 and 2004, but astonishingly a slowing down from 2005 on.

Measures taken to counteract the negative impact of the crisis

National tourist authorities and tourist boards, as well as the tourism industry itself - at local and national levels - have reacted in a number of ways. One way, for example, has been to make their destinations more attractive to customers. In many cases, special committees were set up to deal with the situation after September 11th. Partnerships were established between the private sector and public authorities to tackle problems of financial bottlenecks, e.g. the funding of marketing campaigns or the prevention of bankruptcies in the sector. The measures taken range from price reductions introduced by individual enterprises in the areas of travel, hospitality, tour operating, culture and recreation, to large-scale advertising campaigns by tourist authorities. These have often depended upon substantial injections of finance by both the industry and governments.

Advertising campaigns strongly address the domestic tourism markets. For instance, a campaign run by the Travel Industry Association of America (TIA) features the slogan "See America." A comprehensive plan of action was developed by the Travel Industry Association of America to enable the industry and government to work together to reassure the public that travel is safe while rebuilding consumer confidence in travelling both to and within the United States. The USA Small Business Association (SBA) approved a total of US$21.6 million on disaster loans for small travel agencies. The average loan size was US$50,579. One of the reasons was that SBA recognized the hardships facing travel agencies nationwide in the rapidly changing travel market, especially after September 11th. The SBA also recognized that travel agencies were dramatically affected by reduced airline commissions and advances in technology, specifically the use of the Internet.

A campaign by the Canadian Tourism Commission (CTC) invited Canadians and near-border Americans to re-discover or discover Canada. In fact, the Canadian Government has been supporting a massive marketing campaign by contributing the equivalent of US$13 million to help the hotel industry recover from a loss of US$3.5 billion incurred in 2001. Austria also ran a major campaign to attract European tourists to their winter sports resorts aimed at replacing missing overseas visitors. One of the most impressive programmes to boost tourism is being developed in the United Kingdom. To overcome the country's deep crisis in international tourism, the British Government and a number of British tourist industry organizations have announced they will back a new international campaign, boosting marketing support for the British tourist industry to the tune of £ 40 million. The government will provide up to £20 million for this new initiative addressed to seven main market countries and aiming to recover at least half of the estimated £2 billion lost by the tourism industry last year.

The campaign features over 1,000 special offers put together by both big and small companies throughout the industry, giving discounts of sometimes as much as 50 per cent. The New Zealand government announced that it is to make US$2.5 million available for the expansion of New Zealand's official tourism classification and-quality-grading company, Qualmark, in order to help attract more tourists through improved quality of services. Another example of government support is that of India where, among other relief measures implemented, a 10 per cent expenditure tax on all food and beverages served by hotels was lifted in federal budget announcements in the spring if this year.

In Japan, the government is providing financial support to the troubled travel industry, of which 99 per cent consists of small and medium-sized enterprises. The government's credit association is doubling the maximum loan guarantee for tourism enterprises and government-affiliated financial institutions are doubling the ceiling for unsecured loans. In general, the tourism sector has been given increased attention by public authorities as a result of the 2001-2002 crisis. Even in southern Europe, which did comparatively well immediately after the September 11th, both State authorities and the public were sensitized by the crisis and initiatives for privatepublic partnership increased. One example is a special Euro-Mediterranean group that has been set up on the initiative of Spain and Tunisia to encourage tourism cooperation between the countries of the European Union and Northern Africa.

The industry's endeavours to cut costs, however, strongly affect employment. For example, in a ranking of the 12 types of measures practised most frequently by the US lodging industry to reduce costs, staff reductions and the reduction of working hours are at the top of the list. Six of the first seven measures listed affect staff directly, not to mention any indirect effects from other business reductions. It is not just ordinary staff but also managers that are losing benefits. Management incentives are predicted to reach only 0.3 per cent of revenues in 2002, or 80.6 per cent less than in 2000, the peak year.

Cross-Cultural Challenges in Hospitality Industry

Since the 1980s hospitality industry and academia have shown growing interest in employee empowerment. In hospitality empowerment is seen as the notion of devolving decision-making authority and responsibility to frontline employees for control and enhancement of service quality and customer satisfaction during service delivery. Despite the inherent international character of hospitality operations empowerment theory has only sporadically addressed the crucial question of whether the global propositions of the "empowerment doctrine" are relevant in different cultural contexts.

As a result, advocates of empowerment propose radical changes in the form of organisation (e.g. hierarchical structure) and in the management style and make arbitrary assertions that employees will invariably embrace empowerment. However, according to Hofstede's work on national cultures the form of management style and organisation and even what constitutes a supreme motivator are to a significant degree specific to the cultural context of each country. A handful of authors who have addressed the notion of empowerment from a cross-cultural perspective clearly challenge the universal validity of the propositions of the empowerment doctrine.

In social anthropology it is more or less agreed that the core of culture is formed by deeply embedded values, which are shared by the members of large (sub-) populations. In an analogy of the way computers are programmed, values are instilled in individuals implicitly and unconsciously in early childhood, in a similar way that software is installed on computers. Extending this analogy with the programming of computers, Hofstede defines culture as "the collective programming of the mind which distinguishes the members of one group or category of people from another".

The theory of Hofstede on national cultures resulted from an international survey conducted between 1967 and 1973 among the employees in the subsidiaries of IBM in 72 countries. The research resulted in the identification of four dimensions of national cultures, which "point to four basic problems every society has to solve in its own way: inequality, togetherness, gender roles, and dealing with the unknown". These were named respectively "power distance", "individualism-collectivism", "masculinity-femininity" and "uncertainty avoidance".

The implications that these dimensions raise for, among others, the role pair of boss-subordinate in the workplace point to power distance and uncertainty avoidance as the ones that more clearly influence the application of empowerment. Power distance is defined as "the extent to which the less powerful members of institutions and organisations within a country expect and accept that power is distributed unequally". Power distance reflects, among others, the degree to which individuals in various cultures prefer to be managed in a directive or a participative style. Uncertainty avoidance is defined as "the extent to which the members of a culture feel threatened by uncertain or unknown situations" and reflects individuals' emotional need for rules and formal structure.

In what follows, the specific propositions of the empowerment doctrine about leadership, organisation and motivation and their cultural relativity on the basis of Hofstede's dimensions of national culture are elaborated. The central proposition in the notion of empowerment is that decision-making authority should be delegated to frontline employees, so as to enable them to respond immediately to customer requests, problems and needs. What appears to be at first glance merely a different approach of managing the service encounter, in essence represents a drastic change in the work roles of both managers and subordinates.

The former are required to shift from commanding and controlling to delegating and trusting and the latter from following instructions and rules to making their own decisions and even breaking the rules if this is necessary. In short, the empowerment doctrine promotes a shift in the leadership style and in particular the abandonment of traditional autocratic and directive leadership in favour of democratic and participative leadership. However, the proposed shift in the leadership philosophy may not be possible in large power distance cultures, where "superiors and subordinates consider each other as existentially unequal". This "existential inequality" is expressed in analogous inequalities in power between managers and subordinates and a clear distinction of their roles.

Employees will expect to be told what to do and "will be less inclined to question authority, demand to be consulted or act independently from their superiors". The empowerment proposition for such independent thinking and acting would place a demand on them "to, at least, suspend their beliefs about the 'sanctity' of chain of command and authority". In such cultures "the ideal boss, in the subordinates eyes, is a benevolent autocrat or 'good father'". Managers are, hence, expected to lead in a directive way and "it would not be acceptable within such a culture for the leader to share or delegate the burden of decision-making to his followers".

Employees would tend to lose respect for managers who do not have precise answers to their questions and seek their input in the decision-making process. Failure to meet this demand for leadership and direction on behalf of the manager may even be considered as "negation of leadership rather than a positive step towards empowerment". But even in the occasion that employees would welcome such radical change in their work-roles, in practice it would take considerable time and effort before they engaged in independent thinking and acting, which "in their experience typically results in failure and reprimands for breaking the rules or rocking the boat". In conclusion, the most principal proposition of the empowerment theory is unlikely to be realised in large power distance cultures.

The empowerment theory also proposes radical changes regarding organisational structure and the centralisation of power and formalisation of processes. In particular, empowerment demands a flattened or delayered hierarchical structure with few intermediate management and supervisory layers in order to allow organisations to become flexible and adaptive to rapidly changing environmental conditions. Such flattening of the organisational pyramid further connotes - or results in - decentralisation of power and the dispersion of decision-making authority throughout out the organisation.

The empowerment doctrine supports also an environment of low formalisation with less policy manuals or detailed procedures and rules, so as to allow employees the latitude and flexibility to exercise discretion and initiative. The flexible, horizontal organisation systems proposed by the empowerment doctrine are not necessarily suitable for large power distance and strong uncertainty avoidance countries. In large power distance cultures the "existential inequality" between superiors and subordinates is reflected in tall hierarchies with many intermediate layers of supervision.

The decentralisation of power requires both superiors and subordinates to suspend their deeply embedded values of their existential inequality and their distinct roles. In practical terms, superiors, on the one hand, would need to give up - along with power - the status and privileges attached to it (e.g. the right to conform to different laws and rules), which they have patiently acquired through the years (i.e. progression through seniority).

On the other hand, subordinates would be required to disregard their strong dependence on their superiors and function independently with the commensurate degree of risk and uncertainty that this engenders. The latter would be a highly undesirable development in strong uncertainty avoiding cultures, where uncertainty and ambiguity are the very situations that individuals strive to avoid.

The proposition of the empowerment theory to replace rigorous control of employee behaviour through instructions, policies and orders with trust is also unlikely to be fulfilled universally. In strong uncertainty avoiding countries the existence of many formal and informal laws and rules controlling the work process satisfies an emotional need of people for structure and order. Employees in such cultures feel comfortable in structured environments, whose establishment is, in fact, an attempt to prevent uncertainty and ambiguity in the conduct of people.

Managers are expected to issue clear instructions and subordinates are expected to follow these instructions and to adhere to the formal and informal rules of the organisation. Therefore, imposing a culture of low formalisation in fulfilment of the demands of empowerment would introduce an environment of high uncertainty and ambiguity, in which neither managers nor subordinates would be willing or able to function.

In conclusion, the propositions of the empowerment doctrine concerning the form of organisations are unlikely to be realised in large power distance and strong uncertainty avoiding cultures. Probably the strongest evidence of the inherent cultural bias of the empowerment doctrine is the explicitly stated common belief that employees will invariably embrace the notion "in the prospect of increased personal autonomy, development and job satisfaction". Lashley, for example, claims that empowerment can "tap into the enthusiasm and intrinsic sources of job satisfaction" thanks to increased variety and responsibility introduced in the job content and greater feelings of involvement and importance. The alleged virtuous circle is completed with greater employee commitment resulting in reduced labour turnover and with improved customer service.

Evident in the above claims are the individualist values prevailing in the American and Anglo-Saxon cultures, which, according to Hofstede, promote the fulfilment of "self-interest" as the ultimate motivator of human behaviour. Clearly the allure of the empowerment doctrine is based on this fundamental assumption, which has been expressed in the most popular theories of motivation developed in the United States. Empowerment, in fact, is argued to build on the higher needs for self-actualisation and achievement, which are the top motivators in Maslow's Hierarchy of Needs and in Herzberg's Two-Factor Theory.

Hofstede argues that in strong uncertainty avoiding cultures the need for safety or security prevails over the need for achievement and, hence, individuals are not likely to find any satisfaction in - and to be motivated by - the increased opportunity for self-development and challenge offered by empowerment. In addition, the inner need for structure in such cultures and the unwillingness to take unfamiliar risks is likely to deprive individuals of any satisfaction that "empowered" jobs can offer in terms of personal autonomy, responsibility and self-direction.

On the contrary, empowerment is likely to appeal to individuals in weak uncertainty avoidance cultures, where individuals are motivated by the prospect of selfdevelopment and growth. Also, feeling comfortable with uncertainty and less structure, they are likely to enjoy and make use of the opportunities that empowerment offers for responsibility, autonomy and the exercise of discretion. In sum, the empowerment doctrine is likely to be more appealing to individuals in weak uncertainty avoiding cultures and less so to individuals from cultures at the opposite pole.

The validity of the above substantiated considerations about the "cultural relativity" of empowerment was examined through a study conducted among the employees and managers of 16 upscale and luxury hotels of a single well-known American multinational corporation. The selection of a single company was deliberate, in order to create well-matched and functionally equivalent samples by holding constant other salient characteristics, such as industrial and organisational culture, occupations and so on.

The study was conducted in seven European countries, namely England, Sweden, the Netherlands, Italy, Greece, Portugal and Belgium. Data were collected by means of both qualitative and quantitative methods. Qualitative methods involved 45 semistructured interviews with managers of the hotels, usually with the Human Resources Manager, the Front Office Manager and the F&B Manager and one interview at the company Head Office in Brussels. The interviews aimed at generating information regarding the human resources policy of the hotels, including employee empowerment, and the company in general and were in most cases recorded on tape.

Quantitative methods involved a survey among 427 customer-contact employees, working in the Front Office and F&B departments. The survey measured quantitatively employees' values, as well as their perceptions on management practices (e.g. rewards, information and communication) and management style. The survey also measured employees' self-reported behaviour during service delivery. Quantitative data were analysed by means of multiple regression analysis. Due to the low sample size at the country level, the sample was divided in two cultural clusters on the basis of theories on taxonomy of cultures, namely a south European cluster, including Italy, Greece, Portugal and Belgium and a north European cluster, including England, Sweden and the Netherlands.

The findings of both the qualitative and the quantitative research uncovered a clear difference in the empowerment process between hotels in north and south Europe. The evidence from the employee survey revealed that employees in hotels in the north are more inclined to demonstrate extra-role "empowered" behaviour during service delivery (e.g. creative rulebreaking to please customers, "complaint ownership", "going out of their way" to please customers). The information provided from the interviews supports this finding, as managers in the north generally reported that frontline employees in their hotels consistently engage in "empowered behaviour" and usually would not resort immediately to management assistance when dealing with non-routine service situations.

On the contrary, in the south managers generally reported that employees do not engage in behaviours outside the boundaries of their role and would usually seek management involvement when faced with non-routine situations. Especially clear was the discrepancy between north and south when it comes to making financial concessions to guests. While in the north managers generally stated that employees would often make financial concessions even without consulting first with a manager, in the south managers were in most cases very clear that making financial concessions is not part of the job of employees.

In addition, the survey showed that employees in the south perceive they have a lower degree of authority to function independently in their job than their colleagues in the north. This finding was reinforced by the information provided by the managers. Especially in Italy, which comprises the largest proportion of the sample in southern Europe and to a slightly lesser degree in Belgium, managers made a clear distinction between the extent of authority and responsibility that should be - and eventually is - allocated to the managerial, supervisory and line level.

In Italy, it was frequently stated that this distinction between the roles of managers and subordinates and their commensurate level of authority is prescribed by union agreements, which means that this absence of delegation is also institutionalised. On the contrary, in the north, managers in most cases reported delegating an increased degree of authority to employees and readily identified employees' "expertise" in detecting customer needs and problems and in selecting the appropriate means to ensure customer satisfaction.

Clear differences between hotels in the north and south Europe were found in a number of Human Resource Management practices that were investigated. Both the qualitative and quantitative data revealed a discrepancy between the north and the south in relation to openness of communication and sharing of information. Employees in the south perceive that communication in their working environments is less open and that management shares less information with them compared to employees in the north.

The information provided by managers confirms these results. Managers in the north reported a range of formal processes facilitating especially upward communication and a variety of means for the devolution of organisation-related information, including new technology (intranet). In contrast, most managers in the south did not report formal processes of upward communication and expressed a general preference for an informal communication climate. Occasionally, they even expressed their reluctance to devolve sensitive information to lower level employees and obviously acknowledged the absence of formal channels to do so.

A discrepancy exists between north and south in terms of the extent to which rewards are linked to performance. The qualitative data reveal a greater availability of individual incentive schemes in the north and a greater emphasis in the south on equal compensation of employees of the same rank, position and seniority. According to managers in the south it is actually the legislative framework and union agreements that impose equal compensation. Interestingly, this discrepancy was not uncovered from the employee survey, which may pointed that employee rewards for frontline staff in hospitality are not primarily based on individual performance.

Challenges Facing Hotel Managers

American Express has conducted a survey amongst members

of the British Hospitality Association and The Master Innholders, which revealed that staff retention and skills shortages are the biggest obstacles facing today's Gen
number one by over half of responde
reveals that this is an increasing challenge for General Managers, as 82% of those surveyed feel this issue holds greater relevance compared to three years ago.

Top 5 challenges facing general managers:

— *Staff retention / skills shortages*: 96% of respondents voted this issue as one of the top 5 current obstacles

— *Increasing legislation*: 72% ranked increasing legislation as a top 5 challenge, with 86% feeling this is more pressing than in previous years.

— *Business cycle*: 72% believe that the cycle of negative economic conditions continues to be a key factor in 2004, with equal relevance to conditions in 2001.

— *More demanding owners*: 45% ranked this as a top 5 challenge, with over half (55%) considering this issue has the same relevance as it did three years ago.

— *Impact of 'e' distribution channels*: 43% feel this will affect the industry this year, and the majority (86%) feel this is considerably more relevant than three years ago.

While airlines have had to respond to recent world events by heightening security, only 2% of respondents ranked hotel security as their number one challenge. Hans Lindh, Head of Hotel & Restaurant Industry, American Express Establishment Services says, "Staff retention and skills shortages have always been major issues for hotel management. However as our research shows, this challenge has fast risen up the agenda in recent years, reflecting a continued need to improve the industry's image as an attractive employer and as a career. Industry initiatives like Springboard will therefore continue to play a key role in attracting people to the industry, as will increased focus on leadership and staff training to aid retention."

The following table shows a breakdown of the percentage of respondents ranking what they view as the top five most important challenges for General Managers:

Top 5 competencies required from General Managers:

— *People Leadership*: 100% of survey respondents voted people leadership skills in their top 5 desired competencies in a General Manager. The majority (69%) also felt that this has more become important than in recent years.

— *Business Acumen*: A significant 84% of those surveyed ranked business acumen amongst their top 5 competencies required in General Managers.

— *Knowledge of sales and marketing*: 82% ranked sales and marketing expertise as a vital requirement - with over half (58%) saying it has increased in importance compared to three years ago.

— *Financial knowledge*: Over two thirds (68%) of hotels desire strong financial knowledge in General Managers. While they feel it is just as relevant today as three years ago, they do not think it has grown in importance over the past few years. Fifty-one percent think it has the same relevance, as a competency for a GM, than three years ago.

— *Operational knowledge*: Only two thirds (64%) of respondents ranked good operational knowledge as a top 5 competency - although only 15% feel it has moved up the ranks of importance in the last three years. The majority (67%) feel it has just the same relevance.

NEW THREATS

The combined effect of a prolonged lack of economic dynamism in the world; safety concerns in view of recent security events, and hostilities in the Middle East; and the Severe Acute Respiratory Syndrome (SARS) have caused a new downturn in the travel and tourism industry that started in March 2003. The impact of SARS, if it is not contained soon, has the potential to profoundly change the lifestyles of large populations far beyond their travel and tourism behavior. It might affect all of the world's service industries, in particular, those which involve the ability of people to get close to each other such as on public transport and in shopping areas.

Since methods of transmitting the virus responsible for SARS are not completely known yet, people may start to avoid non-essential activities that involve personal contact with others. Traveling, shopping and dining out are being replaced by telephone or internet contacts which include orders for home delivery of grocery and prepared food. The impact on the economy could be a multiple of the effect that SARS presently has on tourism, as it would affect residents alike.

In particular, the impact on employment would be devastating, as large portions of the services sectors affected are in the hands of labour intensive small and medium sized enterprises. The worldwide travel and tourism industry is threatened by a crisis much deeper than the slump it has suffered in 2001-2002. If the situation does not improve soon, it is estimated that by May 2003, no less than 5 million jobs may be lost in the worldwide travel and tourism industry. This represents over 6 per cent of the industry's total employment. If these losses materialize, they will add to the six and a half million jobs which were previously estimated to be lost as a result of the crisis in the travel and tourism industry in 2001-2002.

Taken together, all job losses from the crisis of the travel and tourism industry has been suffering since 2001 would represent 14 per cent of all jobs in the industry or one in seven. The travel and tourism industry represents over 4 per cent of GDP and over 3 per cent of employment worldwide. If indirect effects are calculated in, its portion rises as high as 11 and 8 per cent respectively. Direct employment in the industry's enterprises already amounts to about 80 million jobs.

In certain tourism-oriented countries, the travel and tourism industry accounts for more than ten per cent of direct jobs, such as Spain or the Caribbean. It is certainly the most important employer worldwide after agriculture and public services. The travel and tourism industry and its suppliers together are estimated to provide close to 200 million jobs globally, or one in every thirteen. As a rule of thumb, one direct job in the travel and tourism industry generates employment in up to two jobs for the industry's suppliers.

The slump comes after a strong global economy had propelled an international tourism boom in the late 1990's. In the peak year 2000 alone, the number of international trips increased by a record 4.5 per cent from the year before. In the year 2001 the industry suffered disastrous effects from the events of 11 September and declined by around 1 per cent on an annual basis. Although tourism demand has traditionally bounced back one year after any crisis motivated by security problems, it did not grow significantly, if at all, during 2002. This is measured by the performance data of the hotel industry, tour operators and airlines in spite of an increase in traveling across international borders recorded by the World Tourism Organization (WTO).

In early 2003, the travel and tourism industry developed some optimism. after it had been suffering an economic downturn which started in 2001 and worsened after the events of 11 September that year. It grew very little during 2002 and was stagnate in the last quarter due to security incidents in Djerba, Bali and Mombasa. The optimism dampened as early as March by a general reluctance to travel in view of the hostilities in the Middle East. The Asian and Pacific region experienced few effects until it was exposed to new fears coming from a new disease.

Business travel, which is suffering the most, generates far more than half of the hotel industry's profits as business travelers roughly spend twice as much money compared to leisure tourists. The sub-sector has been suffering severely since 2001 due to the general economic downturn. More and more companies are now deferring business travel and are relying on supplementary methods of communication such as video-conferencing and other electronic means.

New habits of communication are feared to eventually replace traveling even when things return to normal. This would have a devastating long-term effect on the hotel and airline industries, as video conferencing could cost airlines between 5 and 10 per cent of their bookings over the next 18 months. The recently booming "meeting, incentive, conference and exhibition" business, known as MICE, is the main victim. Revenues from business travel in Asia represent a fifth of the $250 billion global business travel industry and will fall by about 5% this year. That compares with double-digit growth in each of the past four years.

Three categories of locations can be distinguished by the degrees of impact in accordance with either being affected directly or indirectly by SARS, or by the general economic and political insecurity and continuously weak economy. Countries significantly impacted on by SARS are China; Hong Kong, China; Singapore; Taiwan, China; and Viet Nam. The second category includes the rest of South East Asia and the Pacific while the rest of the world with a few exceptions makes up the third .

Rough estimates indicate that since March 2003 the travel and tourism business lost at least one third of its activities in the locations directly affected by SARS. These are Hong Kong, China; the Guangdong province of China; Taiwan, China; Singapore; and Viet Nam. In other South East Asian countries, the immediate loss is about one third, but may be about 15 per cent in the long run. In most of the rest of the world the industry is estimated to lose about 5 per cent this year instead of growing at the rate of one or two per cent expected earlier.

The Asian and Pacific region suffered less from the crisis in tourism of the years 2001 and 2002 than other parts of the world. Growth was largely due to fast growing intra-regional tourism from markets such as Indonesia, Malaysia, Singapore and the booming Chinese market, which also comprises Hong Kong, China. Even after the security incident in Bali last October, the impact seemed to affect international travel to Bali only, especially from Australia.

Bali partly managed to fill the gap by attracting local budget travelers from Indonesia and neighboring countries. Amongst Indonesia's annual 5 million foreign visitors, 4 million are from the ASEAN countries, Japan, the Republic of Korea and China, leaving only 20 per cent of international arrivals to Australia, New Zealand and long haul tourism. Domestic tourism is increasing in Indonesia, but international visitors spend on average 12 times more money during a trip.

Intra-regional tourism benefited the industry with two-digit annual growth rates in January and February regardless of the delicate geopolitical situation. Prospects were good for the year, as the impact of Middle East violence on Asian and Pacific tourism was not expected to affect the region in its peak season June-September. However since March 2003, the situation has dramatically changed due to SARS. Long haul travel activities are being particularly affected. Some U.S. carriers have reported that overseas bookings have plummeted 30% to 40% in April from yearago levels.

Traffic to Asia in particular dropped by 25 per cent since December which was mainly due to SARS as it was only off about 5 percent in early March. BA carried 27 per cent fewer people on its scheduled services to the Asia-Pacific region in April than the same month last year. Some hotels and airlines serving the directly affected locations are down to no more than 30 per cent of their usual business. A bus company in Singapore reported that business is down to ten per cent of their usual and an outbound travel agency in Taiwan, China, reported 80 per cent cancellations. Some shopkeepers in Hong Kong, China, reported not seeing a single customer for several days.

In countries directly affected, not including China, the travel and tourism industry provides about one million jobs, representing between 2.0 and 3.5 per cent of the total number of jobs. Considering all employment dependent on travel and tourism, the total is about three million jobs or between 5.4 and 9.9 per cent according to the country. China is a special case due to its size. Whilst only one province is affected directly, the travel and tourism industry of the whole country may be affected soon. Over 15 million jobs or 2.1 per cent of all jobs are provided by China's travel and tourism industry directly, and as many as 52 million or 7.2 per cent of total employment including indirect jobs.

The core countries affected by SARS have been suffering from high ratios of cancellations in leisure travel as well as in business travel and congresses. Tourist arrivals to Hong Kong, China, slumped by some 30 per cent in the last week of March alone, and likewise, 30 per cent of flights from there were cancelled in March and early April. The occupancy rate at some of the most fashionable hotels in Hong Kong, China, was down to just 14 per cent reportedly due to the fact people could not go home. In general, the hotel occupancy in Hong Kong, China, dropped to single-digit rates. It is estimated that Hong Kong, China, will continue to lose 6 per cent of GDP as long as the outbreak lasts.

According to IATA, airlines have reduced a large percentage of flights to the region as tourist bookings have fallen by about 80 per cent impacting overall air travel to areas such as Hong Kong, China, by 30 to 40 percent. Regional airlines still in the process of recovering from the 2001 slump, have been among the first to send out an SOS, reporting that up to 50 per cent of their flights have been suspended. Cathay Pacific (Hong Kong, China) warned that it may have to ground its entire fleet with passenger numbers predicted to fall to 20 percent of normal levels by May 2003. In general, the hotel occupancy rates in the affected locations are estimated to be down by over 30 per cent and are estimated to be below the break-even point which obliges the management to cut costs drastically.

In Vietnam, hotel occupation dropped from as high as 98 per cent to a level of only 40 per cent. In the longer term, hotel occupation rates in the affected locations may stabilize with a loss of 15 per cent. Hong Kong, China, and Singapore were amongst the poorest hotel performers in March as compared to the same period the year before, with revenue per available room falling 16.8 percent and 15.4 percent respectively (measured in US dollars). Hotels across Asia experienced a 4.6 percent decline.

The Singapore Tourism Board's (STB) latest weekly report showed that visitor arrivals to Singapore plunged by 61 per cent in the first thirteen days of April, as compared to the same period a year ago. Hotel occupancy rates fell to 20-30 per cent from normal levels of at least 70 per cent. The Singapore Retailers Association reported that sales could fall by as much as 75 per cent for some firms. Declines for general stores perhaps range between 20 and 40 per cent. Singapore, which like Hong Kong, China, relies heavily on tourism and its role as a regional transport hub, has cut its economic growth target to 0.5-2.5 per cent this year from 2.0-5.0 per cent due to the impact of SARS.

The late disclosure by the Chinese authorities of the SARS outbreak has created additional concern amongst travelers. The Grand Hyatt in Shanghai reports an occupancy rate of about 30 percent compared to last year in mid-April when the hotel was nearly full. Four major hotels in China's biggest city were closing for up to three months due to a lack of guests. Prospects are that the tourism business in all of China may be affected soon. For example, Beijing and the Shanxi province came to the limelight when the World Health Organization recommended the postponement of unnecessary travel there, as the organization had done earlier for Guangdong province and Hong Kong, China, and later also for the provinces of Tianjin and Inner Mongolia (as well as Taipei).

China's inbound tourism industry brought in 20.4 billion dollars in foreign currency last year. Chinese tourism industry analysts estimate that tourist income may fall at least 40 percent this year. In other Southern and South Eastern countries the travel and tourism industry is affected by SARS because of direct flight connections to indexed locations. These countries are also affected by the notoriously weak geographical knowledge on the world's tourism markets which has resulted in a large de-facto "quarantine area" in South East Asia and Oceania comprising important tourism countries such as Thailand, Indonesia, and the Philippines as well as neighboring Australia and New Zealand.

The travel and tourism industry in Indonesia, Philippines, and Thailand alone means over 5 million direct jobs and a total of over 12 million jobs including those in other industries depending on travel and tourism. Their share in total employment is as high as 11 and 12 percent in Thailand and the Philippines respectively, i.e. one in nine jobs is in danger when travel and tourism slumps, and 6 per cent in Indonesia. Direct employment in the industry represents as much as 2.4 percent of total employment in Indonesia, 4.1 and 5.7 per cent in the Philippines and Thailand respectively.

In Indonesia, foreign arrivals are down about 15 percent for Jakarta and 20 percent for Bali airport. Flights arriving from Singapore fell by about 30 per cent due to the outbreak of SARS in that country. Indonesian hoteliers have seen occupancy rates quickly tumble by about 10 per cent. Mariott reported from Jakarta that its luxury hotel there now posts a mere 40-50 per cent occupancy rate, one of the city's best. Normal hotel occupancy rates are between 60 and 70 per cent. The decline therefore represents at least 15 per cent.

Thailand, one of Asia's most tourist-dependent economies, earned over USD 8 billion from tourism last year. Thai Hotels are currently experiencing cancellation rates of 5 to 10 per cent. According to the Tourism Authority of Thailand, which predicted a 40 per cent drop in foreign visitors coming to Thailand between April and June, About 50,000 visits were cancelled for the period between April and June 2003. Five-star hotels in Bangkok are expecting an occupation rate of no more than 30 per cent for April 2003. April falls in the lean season, but in April 2002, the occupancy rate was still 50 per cent.

Arrivals to Cambodia fell by 22 per cent in March and are frustrating expectations to reach the one-million line this year, which would have brought a 25 per cent increase over last year's 800,000 visitors. Cambodia had reported a 17% one-year increase in arrivals in January and 33% in February. The Philippines remained moderately affected so far, with a drop of foreign arrivals by 4 percent in the first quarter as compared to the first quarter 2002, in spite of a 13 per cent growth in January over last December. Similar to other visitors, South East Asian finance ministers postponed their trip to Manila where their annual summit was to be held, although the ASEAN labour ministers' meeting took place in West Nusa Tenggara (Indonesia).

Malaysia, on the other hand, is heavily affected, with airline bookings down by 40 per cent and hotel occupancy rates by as much as 30 per cent. Industry officials say some hotels are seeing occupancy rates dip as low as 30 per cent after cancellations not only from tourists but also due to postponements of corporate events and shows. In Korea, arrivals of foreign air travelers dropped by almost 50 per cent last month compared to April 2002.

Australia and New Zealand are affected by SARS, as far as the markets on the northern hemisphere are concerned, due to their geographical closeness to the core countries hit by SARS. As both countries are considering closing their borders for travelers from those locations, arrivals through places like Singapore or Hong Kong, China, would also be excluded. Already at present, the Australian Hotels Association has confirmed bookings are on the decline with occupancy falling by between 20 and 30 per cent and tour operators informed bookings of foreign visitors for May are down 30 per cent and for June 20 per cent as compared to the same period last year.

In February, cancellations were highest from the United States, the United Kingdom, Italy, France, Germany and the Netherlands, but since news of the SARS virus hit the headlines, Asian tourists have also started canceling. SARS is expected to cost New Zealand about 20,000 visits by East Asian tourists and around $60 million a month at its present level. The Tourism Board predicts a 15 per cent decline in the number of Asian visitors in 2003 because of the SARS infection scare.

In other parts of the world, the 2001-02 crisis of the travel and tourism industry continues due to SARS, but hostilities in the Middle East and the state of the overall economy play the predominant role. Instead of growing by between 1 and 2 percent or more as forecasted by several sources, the worldwide travel and tourism industry is facing cancellations and a low level of reservations for the northern summer, as well as a continuous decline of business travel. The loss of income is estimated to amount in average to 5 per cent for the world without South East Asia, China and Oceania.

The average current loss of income for the whole world is estimated to amount to 7 per cent. In the United States, hotel occupancy rates have dropped to an average of 60 per cent in the first quarter 2003 from 63 per cent over the last year according to the Travel Industry Association. They are forecast to run at about 59 per cent occupancy rate this year, whilst 65 per cent is usual in good years. As the break-even point is as low as about 55 per cent, the hotel sector in general in the United States continues being profitable due to rigorous cost cuts in the recent past and a rational financial structure. However, PricewaterhouseCoopers projects a 7 per cent drop in the sector's earnings this year to $15 billion from $16.1 billion in 2002.

Hotel loan delinquencies are at their highest level since the early 1990's. The negative outlooks for 2003 follow the release of 2002 data by the United States Department of Commerce indicating that the number of international visitors declined 7 percent compared to 2001. The demand for hotel rooms in the United States has never been weaker than at present; since before the war in Iraq and the development of SARS most companies were only traveling on business trips deemed essential. This is a significant blow to the industry since at least three-quarters of the hotel sector's profits come from business travelers.

The number of business trips this spring is expected to fall 2.5 per cent from the spring of 2002 and 13 per cent from spring 2001, according to the Travel Industry Association. A quick recovery is not likely. The Mexican tourist industry reported a 17.1 per cent slide in reservations since the beginning of the war in Iraq. More than 3.5 million travelers have called off their trips to the country, which calculates into losses of up to USD 700 million in revenues or 9 per cent of the tourism sector's annual hard currency income. International travel to France had rebounded in the past year but is currently levelling down. In Germany, the hotel federation announced that the industry might have to cut as many as 10,000 jobs.

Hotel bookings in Rome, Florence and Venice, Italy, ordinarily throbbing in springtime with art lovers and business travellers, were reported to be down by half compared with last year. In Paris, occupancy rates at four-star hotels dropped 25 percent by March 25. The drop was mainly attributed to the absence of Americans, who accounted last year for 19 per cent of all foreign visitors to Paris. Second to the Japanese at USD 652, Americans usually spend more in Paris, an average of USD 328 a day, compared to USD 133 by British tourists, and USD 78 for Germans. In Turkey, foreign tour operators cancelled between 15 and 40 per cent of reservations. Despite being 1,000 kilometers from Iraq, congresses in Istanbul are being postponed. This affects Turkey's goal to attract 15 million tourists and earn USD 11 billion from tourism this year. There is still hope for tourism to recover as Europeans typically visit Turkey in summer.

Loss of Job

The estimates are rather conservative since they are based on a linear relationship between the loss of economic activities and the loss of jobs in the industry. As a rule of thumb, employment in the travel and tourism sector is constant only under the condition of at least a 2 per cent growth of the industry to keep the demand for labour up against increases in productivity. The absence of normal growth is a factor leading to the reduction of employment, which is not reflected in estimates above.

The present job losses come after the industry already lost an estimated six and a half million jobs before the recent events such as SARS and the hostilities in the Middle East. The compounded loss of eleven and a half million jobs over the 2001-2003 crisis relates to an estimated total of about 80 million jobs in the industry, and represents a loss of one in every 7 jobs. The capacity of the travel and tourism industry to create employment seems to be severely damaged by the recent events.

Additional factors may need to be considered if the slump continues. The longer the loss of travel and tourism activities lasts, the possibility exists that more jobs will be eliminated entirely. The reductions in working hours and manpower tend to become permanent, as the missing labour force is made up for by new working methods such as increased versatility. Remaining core staff is induced to acquire additional skills and also accept more flexible working hours.

A hotel and resort company in the United States and Canada reported having eliminated 15 percent of 38,500 jobs and having trained many of the remaining employees to handle multiple tasks. For example, a front-desk person may also take phone reservations and an administrative person in the back office will help check people in during busy times. Modernization of working methods, working relationships and indeed enterprise structures may thus take a sudden leap forward and leave behind a number of employees with reduced or no income.

The loss of employment affects the socially weaker and lesser skilled staff members, as well as part time workers including high proportions of women, migrant workers and young workers. They risk permanently losing income and jobs as long as the industry does not return to substantial growth rates. The social effect is worse as it severely impacts on those workers who have difficulty finding alternative jobs due to lower skills levels. If indirect effects of the crisis are counted in, that is, jobs which are lost in industries that provide inputs to the travel and tourism industry, the estimated loss amounts to over 13 million jobs worldwide. This is to be added to at least the same impact of indirect effects that was already produced by the crisis since 2001.

As a rule of thumb, one direct job in the travel and tourism industry generates as an indirect effect, two additional jobs in related industries. It should be considered, however, that industries providing inputs to the travel and tourism industry will suffer from downturns in this sector with some time lag only, and employment may not be reduced as quickly as in the sector concerned directly.

Companies of the travel and tourism sector report on measures aiming at rapid cost cutting and in some cases downsizing to reduce over-capacities acquired in the 1990's. In Germany, the world leader in tour operating, TUI, recently announced à USD280 million cost-cutting programme for 2004, and the possible loss of 2,000 jobs. In the course of the crisis in the travel and tourism industry, representative bodies of the industry and government authorities have been trying in many countries to develop measures to mitigate the effects, at least in the short run.

Tourism crisis plans are being developed and government relief measures such as tax breaks are being asked for, although with limited success due to high deficits in government accounts from the economic downturn since 2001. Positive examples of privatepublic partnerships have been developed to promote the tourism destination. In some cases such as Canada, Indonesia, Australia, and others, particular efforts are being made to target closer markets in view of slumping long haul international arrivals.

Tripartite approaches are chosen by government agencies, employers' organizations and trade unions in some countries in line with the recommendations made at the Informal Meeting on the Hotel and Tourism Sector: Social Impact of Events Subsequent to 11 September 2001. Tripartite social dialogue is essential to support affected companies in their efforts to implement temporary cost-cutting measures that will help them stay viable and preserve as many jobs as possible. The measures can include taking leave, working fewer hours, reducing lay-offs to a temporary measure and arranging for workers to improve their skills through government sponsored courses.

A number of companies in the countries affected in South-East Asia and Oceania are asking their workers to take their annual leave or unpaid leave, in lieu of layoffs. In Singapore, the Ministry of Manpower, the National Trades Union Congress and the Singapore National Employers Federation said in a joint statement 15 April that whatever steps necessary should be taken to save jobs in tourism-related industries hit by the SARS scare. The government announced relief measures worth 230 million Singapore dollars (130 million US) on 17 April 2003. The package includes rebates for airport landing fees, a reduction of the levy paid by employers for their foreign workers and property tax relief for shops, hotels and restaurants. It will take effect in May and last until the end of the year. The other most significantly affected economy in the region, Hong Kong, China, is expected to follow suit soon. Taiwan, China, has stepped in to offer support to their local airlines and tourism firms.

6

Yield Management

Waste Reduction and Recycling

Waste reduction includes all actions taken to reduce the amount and/or toxicity of waste. It includes waste prevention, recycling, cornposting and the purchase of products that have recycled content or produce less waste. Waste prevention is one of the easiest methods of reducing waste and can be controlled at the time of purchasing products and services. There is a direct correlation between what a hotel purchases and what it throws away. In addition, all of the resources and economic costs of materials, manufacturing, labor, transportation, packaging, storage and disposal are wasted when a product is discarded.

Cornposting food waste and yard trimmings can save a hotel money on hauling and landfilling fees. However, cornposting is limited by land space at the hotel and/or availability of commercial cornposting sites. Purchasing products made with recycled content creates outlets or markets for the materials collected in the hotel's recycling program. As such, procurement sustains the recycling process and in turn, reduces the amount of waste going to the landfill.

Benefits of waste reduction:

— savings through reduced purchasing costs,

— reduced waste disposal costs,

— reduced energy consumption and pollution,

— conservation of natural resources.

Recycling

Recycling is the process by which materials, otherwise destined for disposal, are collected and used as raw materials for new products. Recycling prevents useful materials from being landfilled or combusted, thus saving energy and natural resources. There are three steps in recycling:

1. collecting the recyclable components of solid waste,
2. processing and manufacturing them into reusable products,
3. purchasing and using the products made with reprocessed materials.

Hotel recycling programs focus on collecting recyclable materials from the hotel's waste stream for recycling and purchasing products made from recycled materials.

Benefits of recycling:

— reduces hotel waste disposal costs by diverting materials from the landfill,

— possibly provides revenues through the sale of recyclable materials,

— demonstrates a hotel's commitment to environmental protection to guests, businesses and employees.

Hotel and motel owners, managers and employees must make a commitment to the hotel's waste reduction and recycling initiatives if these programs are to be successful. To be successful requires initial research in present purchasing and waste handling practices. This includes the cost of disposing of waste, research of products, packaging, vendors, recycling processors equipment and related areas. Further planning is needed in program implementation and employee training. The following actions are suggested to assist in organizing hotels waste reduction initiatives.

General Manager

— Inform appropriate people in the hotel's corporate structure of the program's progress, emphasizing the savings in waste disposal costs, revenue from material sales, the conservation of natural, non-renewable resources and pollution prevention. Seek their commitment and financial support.

— Appoint a waste reduction manager to provide leadership to each of these initiatives.

— Establish an accounting system that reflects monthly waste management costs. A monthly report is needed for tracking waste disposal and recycling information.

— Establish a budget for the recycling initiative. Funds may be needed for purchasing recycling containers, equipment and developing employee motivation programs.

— Assist management and employees in establishing the hotel's waste reduction and recycling goals. Post the goals on bulletin boards so all employees are informed.

— Create employee interest. Involve them in the initial stages of the program so they will take responsibility.

— Incorporate recycling procedures in employee position descriptions. Include recycling procedures in new employee orientation, employee training programs and staff meetings.

— Inform the recycling program manager of planned structural renovations and equipment replacements. These activities may require special arrangements with haulers and materials processors.

— Develop an environmentally green marketing plan for the hotel. Work with the hotel's marketing company to highlight the hotel's waste reduction efforts in all promotional materials.

— Inform employees of the program's progress. Place copies of monthly recycling reports on employee bulletin boards and discuss in employee meetings.

Waste Reduction Manager

The recycling program manager is usually appointed by the hotel's general manager. This position requires an individual with leadership and communications skills who is able to motivate employees. Knowledge of the hotel's present purchasing and waste disposal practices and contract procedures would be helpful. The following actions will assist in assuring success.

— Meet monthly with the hotel's general manager to inform him/her of the program's progress. Present reports of program progress at department managers meetings.

— Work with management to establish a budget for program organization, equipment purchases and employee motivation/rewards.

— Develop recycling procedures to complement the operational practices of the hotel.

— Work with the recycling materials processor to develop a training program on materials preparation and train employees in recycling procedures specific for the hotel and various departments.

— Research recycling container types and equipment prices.

— With the assistance of department managers and employees, design materials collection and storage systems that complement employee work procedures.

— Issue bid invitations to waste haulers and intermediate processors for recycling materials removal/hauling and processing.

— Issue bid invitations for purchasing and leasing recycling containers and equipment.

— Review existing purchasing practices to determine how purchasing decisions impact the hotel's waste disposal costs.

— Develop a monitoring system that addresses employee compliance in the hotel's waste reduction initiative.

— Review guest comment cards monthly. It is important to know how guests perceive the program.

— Inform employees of the program's progress. Seek their ideas and opinions at meetings and in private conversations. Keep them in the loop.

— Evaluate the program's progress often and make adjustments as needed.

No matter how well planned your waste reduction and recycling initiatives may be, they probably won't succeed without the support of the hotel department managers and staff. Successful hotel recycling programs require employee involvement, training and recognition. Involve all employees in the recycling program. They are already disposing of waste as they perform their daily jobs. Recycling is simply disposing of job related waste in a different container from the one used for trash. Seek their ideas in organizing the recycling program and for methods of collecting and storing the materials. Employees are more likely to support a program that they understand and have helped design.

Develop motivational programs that create opportunities for staff involvement and enthusiasm about the recycling program.

Training

— Develop employee training programs that include recycling techniques specific for various departments.

— Include recycling information in new employee orientation programs.

— Inform employees of environmental issues and laws. Employees want to know why the hotel is recycling.

Recognition

— Develop a system of on-the-spot, weekly and monthly incentives for employees who exhibit activity and leadership in recycling and waste reduction efforts.

— Recognize employees and departments that show exceptional involvement in the recycling program.

Hotel guests are becoming more environmentally conscious, They are becoming more inclined to support "green" initiatives in the travel and hospitality industry. Most guests are pleased to see the use of products made from recycled materials, energy efficient light bulbs, low-flow shower heads andrecycling programs in hotels and restaurants that they patronize. Hotel managers report that most guests are willing to participate in the hotel's recycling efforts.

To increase guest involvement in the program, place recycling instructions in guest rooms and locate recycling collection containers in convenient locations in convention halls and public areas throughout the hotel. Always place a trash container along side of the recycling container in public areas to avoid trash being thrown in the recycling bin.

GUIDELINES FOR OFFICES

Waste Reduction guidelines will assist hotel and motel management and employees in reducing the volume and toxicity of waste generated in offices. To successfully reduce waste disposal costs, conduct an audit to determine the types and volumes of waste generated in offices. Plan your waste reduction initiative to target specific products or materials that contribute the greatest volume and/or weight to the waste stream or which create disposal problems. Establish short and long term departmental goals for waste reduction.

The following check list provides a format for your department's audit and additional ideas for waste reduction.

1. *Reduce Paper Waste:*

— Use a centrally-placed or electronic bulletin board for messages instead of making multiple copies of memos.

— Shred office paper and use it to package shipments.

— Eliminate fax cover sheets by using a rubber stamp or sticky notes designed for fax use.

— Reuse envelopes and folders for routing in-house mail and correspondence.

— Include a statement requesting electronic addresses on all written correspondence.

— Electronic mail will save on postage, labor and paper.

— Decrease the number of hard copies of documents by establishing a central filing system to store documents.

— Use electronic mail for correspondence when possible.

— Prepare documents with small margins and double sided printing or copying.

— Make double sided copies when possible.

— Use the back side of paper that is printed on one side for drafts.

— Make scratch pads from paper used on one side by stacking paper and coating one end with rubber cement and cutting in half.

— Edit and proof read documents on computer screens before printing.

— Use staff routing slips when sharing information in-house to stop multiple copies.

— Purge mailing lists to reduce returned mail.

— This saves on paper use, postage and recycling costs.

— Avoid unnecessary print-outs by faxing directly from the computer.

— Reduce unsolicited mail by sending each solicitor a postcard requesting that they remove your hotel's address from their mailing list; reuse their original mailing label for your return address.

2. *Reuse:*

— polystyrene peanuts, shredded newspapers and office paper for packing materials,

— obsolete forms and letterhead for drafts and note pads, file folders, mugs for coffee instead of disposable cups.

3.*Reduce hazardous waste by purchasing:*

— non-toxic inks,

— unbleached and chlorine-free papers,

— nontoxic, biodegradable correction fluid,

— equipment that does not require batteries,

— rechargeable batteries for small calculators communication radios and flashlights.

4.*Support recycling by purchasing:*

— paper products made from recycled paper (i.e., letterhead, envelopes, computer, copier, fax, and laser papers, calculator and cash register tapes, file folders, note pads and legal pads),

— recycled office papers and cardstock for pocket folders, table tents, business cards, and guest cards,

— file boxes, clipboards, binders, and computer diskettes made from recycled materials,

— furniture, carpet and padding, trash cans, and recycling containers made from recycled plastic,

— bulletin boards and floor mats made from recycled tires,

— remanufactured toner cartridges for office machines (i.e., copiers, laser printers, fax machines, typewriters, cash registers, and ATM machines),

— refillable pens and pencils rather than throwaways,

— services from contractors, printers, graphic artists, and other subcontractors who use recycled products and recycle at their businesses.

5. *Other Waste Reduction Strategies*:

— Combine office supply orders; ordering in bulk reduces packaging waste and saves time, energy and money.

— Chart the shelf life of items (i.e., toner, correction fluid) and purchase only the needed quantity.

— Request that vendors avoid excessive boxes, bags and wrappings when shipping to your property.

— Purchase plain paper fax machines. Most documents received on thermal paper are copied and the thermal paper fax is then discarded. Thermal fax paper is not recyclable.

— Organize a recycling program to recycle all office paper including cardboard, magazines, brochures, telephone books and computer papers.

— Recycle typewriter, adding machine and computer ribbons and printer cartridges.

— When you purchase new office equipment, ask about their energy efficiency.

Guidelines for Housekeeping

Waste Reduction Guidelines will assist hotel and motel housekeeping managers in reducing the volume and toxicity of wastes generated in the housekeeping department. To successfully reduce disposal costs, conduct a departmental audit to determine the types and volumes of wastes generated for disposal. Plan your waste reduction initiatives by targeting specific products or materials that contribute the greatest volume and/or weight to the waste stream or which create disposal problems. Establish short and long term department goals for waste reduction and implement appropriate waste reduction strategies. The following check list will provide a format for your department's audit and provide additional ideas for waste reduction.

Reduce Waste

— Select vendors who take back reusable containers, pallets and other waste packaging.

— Request that manufacturers package products in returnable and/or reusable boxes and crates.

— Reuse plastic drums and buckets for recycling or trash collection containers in back of the house service areas.

— Use refillable bath soap, shampoo, hair rinse and hand lotion dispensers for guest rooms to eliminate soap pieces and plastic bottles

— Offer newspapers in centrally located places such as lobbies, restaurants, and news stands rather than distributing them to each guest room to reduce the volume of discarded newspapers and plastic bags.

— Eliminate plastic bags and stickers on newspapers.

— Eliminate the use of plastic liners in ice buckets.

— Refinish and reupholster damaged and dated furniture.

— Remanufacture worn mattresses. Mattress manufacturers will rebuild to the hotel's specifications.

— Replace tissues in bathrooms when they are almost empty or leave new rolls or boxes of tissues for the guest to replace. If company policy is to replace half rolls of toilet tissue, save them and donate to shelters or use in employee restrooms.

— Provide. amenities such as shower caps, shoe wipes on request.

— Use the small concentrate empties for spray bottles for cleaning solutions by adding pumps.

— Redye stained bath towels and wash cloths for reuse as pool towels and cleaning cloths.

- — Redye linens and carpet to match remodeled decor.
- — Rotate draperies to expose different portions to sunlight and extend their useful life.
- — Use reusable baskets or plastic containers for guest laundry, thus avoiding disposable paper or plastic wrap
- — Use clothes hangers left by guests for employee uniforms or return to the dry cleaners and recycle plastic garment bags.
- — Purchase cleaning and laundry products in concentrated form, dilute and pour into reusable containers for individual use.
- — Purchase vacuum cleaners with reusable bags vs. disposables ones.
- — Purchase sheets with a high thread count for longer wear.
- — Provide reusable drinking glasses and coffee cups for guests instead of paper or plastic throwaways.
- — Organize a recycling program for metal, plastic, glass and paper materials.
- — Use pump-style sprays instead of aerosol cans.

Reduce Toxicity

- — Purchase cleaning products with less hazardous constituents.
- — Train employees in proper use of cleaning supplies including the dangers of mixing bleach with other cleaning products.
- — Label all cleaning supply bottles with appropriate labels.
- — Use a hazardous chemical containment pan to prevent leaching of unused chemicals.
- — Check chemical containers periodically for leaks and spills and replace as needed.
- — Store unused portions in original containers in a cool, dry place away from equipment, and keep storage areas well ventilated.

— Dispose of chemicals per label instructions. Call your county environmental protection office if you are unsure of how to dispose of hazardous wastes properly.

— Collect dead alkaline, dry cell and rechargeable batteries used in department radios, flashlights and calculators at a central collection point and dispose of them at the battery recycling area in your county.

— Reduce the need for pesticide applications in guest rooms and employee work areas by removing food containers, garbage and recycling containers daily; clean employee break rooms daily to remove crumbs and spills; and report cracks around doors and windows to the engineering department for repairs to prevent crawling pests from entering

— Check labels on room deodorizers for hazardous chemical content. Deodorizers are often used to mask smoke and mildew odors. Reduce the need for these products by implementing practices, such as changing air conditioning filters every three to four weeks; sealing cracks around windows, doors, and air conditioning units; and operating air conditioning equipment at temperatures recommended by the engineering department.

Support recycling by purchasing:

— Toilet and facial tissues and paper towels made from recycled paper fibers.

— Housekeeping carts, waste containers, caddies, and mop buckets made from recycled plastics.

— Carpet, carpet padding, mats, vertical blinds, and furniture made from recycled materials.

Guidelines for Food and Beverage

Waste Reduction Guidelines will assist hotel and motel food and beverage managers in reducing the volume and toxicity of wastes generated by the food service operations. To successfully reduce waste disposal costs, conduct a departmental audit to determine the types and volumes of wastes generated for disposal. Plan your waste reduction initiatives by targeting specific products or materials that contribute the greatest volume and/or weight to your departments waste stream or which create disposal problems. Establish short and long term departmental goals for waste reduction and implement appropriate waste reduction strategies.

Reduce Waste

— Use reusable metal or nylon coffee filters.

— Use only reusable dishes and flatware in restaurants and employee cafeteria.

— Install a magnet on food waste containers to recover silverware accidently thrown away with food waste.

— Use paper towel dispensers in employee rest rooms that control the amount of paper being dispensed or use less than standard width paper to reduce paper waste as well as litter.

— Use filters in deep fryers to extend the life of cooking oil.

— Distribute condiments, paper napkins, and straws upon request instead of offering them self-serve for take-outs.

— Replace individually wrapped condiment packages with bulk dispensers in restaurants and employee cafeterias.

— Use washable hats and aprons for kitchen employees instead of disposable paper ones.

— Use only cloth napkins, tablecloths and china for restaurants, meetings, and conventions.

— Use stained wash cloths and towels for cleaning rather than paper towels.

— Redye tablecloths and napkins to match new color schemes when remodeling; reuse worn table linens as cleaning rags or remake into aprons, laundry bags, mattress covers.

— Use glass or plastic coasters to reduce cocktail napkin waste.

— Eliminate paper place mats and use cloth ones.

— Reuse five gallon PVC buckets.

— Request fruits, vegetables and meats to be packaged in reusable crates, or recyclable boxes.

Purchase:

— condiments in bulk and dispense the items in smaller, refillable containers.

— concentrate bar mixes; reconstitute and serve from reusable containers.

— dispenser beverages (i.e., juice, iced tea, hot chocolate) in concentrate or bulk and pour into reusable serving containers.

— milk in five (5) gallon dispenser bags delivered in reusable plastic crates, rather than by the gallon. If milk is purchased by the gallon, select recyclable plastic jugs, not waxed cartons.

— frozen foods in bulk and store unserved portions in airtight reusable food containers.

— flours and grains in bulk and store properly in reusable food containers.

— draft beer in returnable/refillable kegs to reduce the quantity of bottles and cans requiring disposal.

— carbonated fountain beverages in refillable cylinders and dispense into reusable glasses.

— beverages with the least distribution packaging.

Reduce Toxicity

— Purchase cleaning products that contain the least amount of hazardous chemicals.

— Use pump-style spray containers for cleaning products.

— Dispose of unused portions of cleaning products according to the labels.

— Dispose of used appliances to scrap dealers who recycle CFC's.

— Purchase refrigerators, freezers, and coolers which are CFC-free.

— Use dry, concentrated dishwasher chemicals in dispensers to reduce chemical spills and waste.

— Reduce the need for pesticide applications by eliminating clutter where pests might hide; sealing cracks around doors where pests may enter; and removing garbage, recycling bins and open food containers at closing time.

— Wipe down all food preparation surfaces, tables and other areas which might contain crumbs to attract pests.

Support recycling by purchasing:

— recycled paper for menus, table tents, guest checks, inventory sheets, calculator and cash register tapes; use the recycling symbol on printed materials to inform guest of your hotel's commitment to the environment.

— paper napkins, place mats, and paper towels are used, purchase ones made from recycled paper.

— serving carts and food storage containers made from recycled plastic and recycle them when they need replacing.

— reusable food storage containers made from recycled materials.

— floor mats made from recycled rubber tires.

— carpet, vertical blinds, and furniture made from recycled materials.

— waste and recycling containers made from recycled plastics.

EcoPurchasing Guidelines for Energy Efficient Equipment

The hotel industry is equipped with energy intensive equipment and lighting. Air conditioning, heating, lighting, ventilation, refrigeration, motors, cooking, laundry, cleaning and office equipment are necessary for hotel operations. Choices made when purchasing equipment can have a major impact on the hotel's operating expenses and on the environment; therefore, it is important that efficiency be considered before purchasing energy intensive equipment.

Air conditioning Efficiency

— Purchase A/C equipment sized for the area of use.

— Look for a high Energy Efficiency Rating (EER); a Seasonal Energy Efficiency Rating (SEER) of 10 or higher is recommended.

— Use high efficiency motors for A/C air handling unit fans (fan speeds).

— Install heat recovery units on large A/C systems to preheat water.

— Compare efficiency of gas and electric air conditioning systems before making a selection.

— Purchase washable, commercial grade A/C filters and change them every four weeks during hothumid months.

Lighting Efficiency

— In purchasing light bulbs, as a general rule, the bulb cost is the smallest cost. Energy costs over a bulb's useful life are many times the original purchase cost. Evaluate bulbs by comparing lumens per watt.

— Convert inefficient incandescent bulbs to efficient fluorescent bulbs in exit signs, guest room lamps and other areas where incandescent bulbs are used.

— The newer 95-watt incandescent bulb replaces the older type 100-watt incandescent bulb.

— Replace incandescent reflector bulbs with lower wattage ellipsoidal reflector bulbs.

— Install lighting controls (timers or occupancy sensors) to turn off lights in unoccupied areas such as storage rooms, employee rest rooms, walk-in coolers, etc.

Increase the energy efficiency of fluorescent lighting systems by purchasing:

— high efficiency fluorescent bulbs; for example, replace a T-8 with a T-12 bulb,

— energy efficient ballasts that provide a more efficient energy transfer,

— reflectors that direct more light out of the fixture into the room,

— high intensity discharge bulbs (Low Pressure and High Pressure Sodium and Metal Halide) for parking lot lighting, security lighting, and for select indoor use,

— photovoltaic lighting systems for lighting advertising signs, walkways, plants, trees, decorative water fountains and remote out-door areas.

Electric Motor Efficiency

— Choose high efficiency motors with variable speed controls.

— Consider the hours the system needs to operate before selecting motor sizes.

— Consult with a qualified energy engineer for a cost-savings analysis; two small otors may operate more efficiently than one large motor.

Kitchen equipment considerations include purchasing:

— air-miser hoods that use outside air as make-up air instead of discharging conditioned inside air,

— a timer for the kitchen exhaust hood that turns off the system when the kitchen is closed,

— smaller convection ovens to replace oversized conventional ovens,

— an air-to-water heat pump water heater for kitchen use,
— refrigeration equipment with high energy efficiency ratings.

Water use considerations:

Water use in hotels varies depending on geographical location, occupancy, property type and season of the year. The average water use is 218 gallons per occupied room per day. Larger properties use more water due to the water intensive facilities often found in the larger hotels, i.e., pools, laundries and restaurants. Excessive use of water increases the cost of operation, increases electrical energy required for pumps, depletes the water supply and puts extra demands on waste water disposal plants.

Increase water use efficiency by purchasing:

— dishwashing machines that use final rinse water for the prewash of the next load of dishes,
— water-saving showerheads and sink faucet aerators: 2.5 gallons per minute (GPM) showerhead and 1.5 GPM aerators are recommended for water savings and guest satisfaction respectively. Showerheads with less water flow often increase the number of guest complaints, a food-thawing machine instead of running water over foods for faster thawing,
— laundry washers which use final rinse water for prewashing of the next load of laundry.

All hotels and motels have sufficient quantities of recyclable materials to justify recycling. Nationally 50-65% of a hotel's waste is recyclable. The amount of recyclables generated in hotels/ motels depends on the following: the type hotel property, occupancy, number and size of meal functions, lounge/bar business, conventions, guest and employee activities and purchasing practices. The type and number of businesses operating in leased spaces in the facility also impact the volume of recyclable materials.

Hotels rent space to a variety of businesses but the most common ones are retail stores, business services and photography operations. Businesses leasing spaces in hotels are often overlooked in recycling programs. Their waste is usually handled by the hotel; therefore, they increase the hotel's waste volume. Consider the amount and type of waste created by a business when negotiating a lease. Cardboard and office paper are the most common and most desirable types of waste created by these businesses. If a tenant produces hazardous waste, the hotel may want to require them to handle their own waste disposal. Hazardous waste usually requires special disposal containers and transportation. Inquire about how they plan to handle it and request proof of proper disposal. If hazardous waste is thrown into the hotel's waste collection or recycling containers, the hotel may be held responsible for disposal problems.

A waste audit identifies the types and approximate quantities of materials for recycling and potential savings due to waste disposal cost reduction. An audit helps identify the size containers needed for collecting and storing the recyclable materials at specific sites in the hotel and at the Central Recycling Center. Although an estimate of the amount of recyclables can be done by direct sorting and weighing of waste, previous hotel recycling programs have proven that an acceptable estimate of recyclables can be made from reviewing purchasing and waste removal records and by evaluating the amount and types of waste generated in each department.

The amount of waste and recyclables produced in a hotel is affected by variables that differ every day. A hotel waste audit, depending on type, produces only information related to the waste and recyclables generated on the day of the audit. The variables mentioned earlier affect the volume of waste on a specific day. These records can be reviewed internally.

Before a hotel recycling program can be implemented, a decision must be made about where the recyclable materials are to be taken for processing. This decision can make the difference between simply disposing of recyclables and receiving money for them. Processors may require different amounts of material preparation required for recycling. This important decision usually involves public or private processing facilities. An intermediate processor collects, consolidates, and prepares quantities of recyclable materials for shipment to manufacturers. Some processing facilities are owned and operated by mills or manufacturers and others are operated by small companies and individuals.

Processors offer different levels of services. Some processors operate buy-back centers where materials are weighed and payment is made for specific materials. Full service processors accept all recyclable materials and may offer other services such as container leasing or loan and materials transportation. A Materials Recovery Facility (MRF) accepts commingled recyclable materials for processing. Most MRFs accept glass, plastic, paper, aluminum and steel materials.

Each operation is different so contact MRF managers for information about acceptable materials and the preparation required. (The hotel should not be required to sort any of the materials, except paper, taken to a MRF). Paper must be kept clean and dry and this is not possible if paper is thrown into collection containers filled with bottles and cans. MRFs may be owned and operated by private industry or by city or county governments. In some states, MRFs are owned by counties and operated by private industry. MRFs may or may not pay for recyclable materials and occasionally there may be sorting or processing fees. Government owned MRFs do not usually pay for recyclable materials.

The majority of the privately owned MRFs do not charge for sorting and pay for some of the materials. Each operation is different so a thorough research is advisable before selecting a MRF. Before choosing a processor for your hotel's recyclable materials, investigate the laws, rules and regulations governing hauling and tipping fees in your area. In some areas recycling may be included in waste hauling franchise agreements with cities and counties. Recycling is exempt from waste franchise contracts in most areas.

The Central Recycling Center

The Central Recycling Center (CRC) is an area outside the hotel where recyclable materials are stored until they are transported to a processor. In larger hotels CRCs may be located in the dock or receiving area.

Consideration should be given to the location of the CRC. For convenience, locate it outside the hotel near an employee service exit. Easy access to the container(s) is needed for vehicles picking up and transporting the materials. The CRC is usually located outside the hotel because of the amount of space required for containers. Odors, insects and rodents are also reasons for locating CRCs outside. The hotel's design, space limitations and vehicle access may make it impossible to locate CRCs conveniently.

Localzoning ordinances may regulate the CRCs location. Visibility of the CRC is a concern for some hotel management. They are concerned with the visibility of recycling equipment and stored materials creating an undesirable appearance for the guests. However, guests are expressing interest in preserving the environment and recycling is one way for the hotel to create a positive environmental image. Well kept CRCs that are visible to guests create a positive image for the hotel. In areas with zoning restrictions CRCs can be concealed with fences, walls or landscaping.

Equipment is needed in the CRC to prepare materials and store them until they are hauled to the recycling materials processor. Equipment usually found in the CRC are storage containers such as a large recycling roll-off, dumpster and/or portable 92-gallon containers. The type materials storage container(s) needed depends on the available space, volume of materials to be recycled and the requirements of the processor.

Hotels with food and beverage operations need a baler for reducing the volume of cardboard created from food and drink packaging. A large trash container for disposing of non-recyclables should be placed in the CRC. If a trash container is not available, non-recyclable materials may be thrown into recycling containers or on the ground.

CRC MATERIALS STORAGE CONTAINERS

Large recycling roll-off containers are the most popular recycling containers for hotel Central Recycling Centers (CRCs). A roll-off container may be compartmentalized for sorted recyclables or simply one large compartment for commingled materials. They are designed with a closed top and conveniently located doors. The top keeps the materials from blowing out of the container when it is moved and provides security for the materials. A locking system is usually optional, but highly recommended to keep out unwanted garbage and recycling material thefts.

Roll-off containers are available in different sizes. To reduce the number of times the container is hauled and/or emptied at the processor, select a roll-off recycling container as large as space permits. If the hotel is paying for each container pull, the savings from recycling is reduced when containers are pulled too often or half full. Recycling roll-off containers are available through manufacturers, waste hauling companies and some material processors and may be purchased or leased. Before making a final decision on storage containers for the hotel's CRC, compare cost benefits of container leasing, purchasing and availability from processors. A worksheet is included in these materials to help in the research of storage containers for your CRC.

Dumpster waste containers are commonly used at smaller hotel/motel properties for storing waste. They are also being used for storing specific recyclable materials such as newspaper, cardboard and office paper. They work satisfactorily for storage of office paper and newspaper. However, dumpsters do not have adequate space to store cardboard or all of the recyclable materials from large hotels. Dumpsters require more frequent emptying, thus increasing the cost of recycling if the hotel is charged for the number of times per week/month the dumpster is emptied.

A recycling dumpster is emptied into a larger container on the back of a truck. The dumpster is not removed from the hotel. Drivers may pick up materials from several locations and dump them into the same container before hauling to the processor or mill. This creates a question as to how the weights are determined. Weights of recyclable materials in dumpsters are usually estimated. Materials are not weighed before dumping into the large container on the truck. If the hotel expects payment for recyclable materials, ask how the weights are determined.

If dumpster type containers are used for collection of recyclables and waste materials at the same location, paint the recycling dumpster a different color and label it. This will help prevent trash/waste materials from being thrown into the recycling container. Dumpster recycling containers may be provided at no cost by some companies or they may be leased or purchased from a waste hauler or intermediate processor.

Large 92-gallon portable rolling containers are the third type of container used in hotel Central Recycling Centers (CRCs) for collecting and storing materials. These containers are convenient for small hotels/motels that recycle only a few items such as aluminum cans and office paper. They are also used at larger properties for collecting and storing newspaper and office papers. Individual containers have limitations on transportation and storage capacity. This smaller container is usually transported by a recycling company or by a hotel vehicle.

Recycling Materials Flow Plans

Once sites for recycling containers have been identified and storage containers selected, establish a materials flow plan-routes to move the recyclables from the temporary storage sites to the Central Recycling Center. Developing a materials flow plan is a very important step in setting up the recycling program. The flow plan establishes the type materials to be recycled at each location, where materials are to be stored, where collection containers are to be used, which employees perform the work tasks and the direction the materials are to be moved through the property. Complete materials flow plan worksheets for every department and mark the directions the materials are to be moved on a property floor plan.

1. The first step in developing the recycling materials flow plan is to review the hotel's floor plan. Locate the storage spaces, hallways, equipment placement and employee traffic patterns in all departments.
2. Identify collection points at work stations and storage areas. From this information a decision can be made on the placement of storage containers and the size containers for each location.
3. Establish people traffic patterns and routes for moving the recyclables through the hotel. The traffic movement routes should be the ones with the least amount of interference with hotel guests and employee activities. Materials flow plans should start at the point where the recyclables are generated and continue until the materials are placed in the recycling storage container(s) in the CRC.

Involve employees in deciding where the recycling collection containers will be located in their work areas, For greater employee participation in the recycling program, locate collection containers in or near work areas where recyclables are generated. This method is called collecting at the source or point of waste generation. For example, place a recycling collection container near the can opener in the kitchen. When cans are opened and emptied they are easily placed in the appropriate recycling materials collection container.

Hotel design, equipment placement and traffic patterns may prevent the containers from being placed in convenient locations in some departments. In this case, place the containers as close to the work area as possible. To reduce the labor costs of recycling, handle the materials as few times as possible. However, in offices and on guest room floors it is necessary to add an intermediate storage center where materials are temporarily stored during the day until they can be moved to the Central Recycling Center (CRC).

After the material collection points have been identified, it is time to choose containers. Match the containers to materials to be collected and to the identified collection and storage spaces throughout the property. Carefully measure the spaces and compare the specifications provided by vendors. Consider color coding the containers for specific recyclable materials in a sorted materials recycling program.

For a successful hotel recycling program, recycling materials collection and preparation must become the responsibility of every employee. The method of materials collection and preparation in the hotel/motel depends on the degree of preparation required by the processor. A small amount of preparation is required by all processors. Preparation of recyclables at the source is easier and less time consuming. As products are used or containers are emptied, place in specific recycling collection containers instead of a trash container. Even if the materials go to a materials recovery facility (MRF), some sorting and preparation is required.

When evaluating how to collect and where to store recyclables, inquire about local fire codes for storing paper and other materials at intermediate and central recycling centers. There are two ways of collecting and storing recyclables in hotels/motels-sorted and commingled. The method is used in your recycling program depends on the requirements of the processor who is handling your hotel's recyclable materials and the amount of labor the hotel is willing to pay for sorting the materials.

In a sorted materials recycling program, materials are sorted by type. Steel, aluminum, plastic, paper and glass are each stored in separate recycling containers. These materials may be sorted further. For example, glass is sorted by color into clear, green and brown.

In a commingled recycling program, all the recyclable materials except paper are stored together. Office paper and cardboard are usually an exception because they must be kept clean and dry, so they are stored separately. The commingled materials are sorted and processed at a MRFs.

Before employees begin collecting materials and placing them in recycling containers, provide training on procedures for preparing materials. Check with your processor as to the degree of preparation and sorting of materials required.

Guidelines for Recycling

The following recycling guidelines provide basic recycling information for materials preparation. Materials preparation may be slightly different with each processor for specific materials.

Paper: Cardboard is the most important material to recycle in a hotel/motel recycling program. All departments in the hotel produce a large volume of cardboard. Proportionately, cardboard is the largest part of a hotel's waste stream. Because of its bulk, cardboard occupies a tremendous amount of space in waste containers and should be removed from the hotel's waste stream, baled and recycled.

Office paper is generated in every department in the hotel. Most of the paper is computer, white ledger, copy, letterhead and envelopes. Incoming mail, magazines and catalogs are additional sources of recyclable paper. Convention attendees and exhibitors leave boxes of printed brochures and presentation materials that are recyclable. Most of these materials are recyclable with mixed paper. Check with your processor about accepting boxed convention materials.

Newspapers add tons to the monthly waste disposal costs of hotels that provide daily papers. Newspapers require no preparation if they are not placed in bags. Collect throughout the property. Place newspaper recycling containers in dining rooms, recreation areas and lobbies for guest participation in newspaper recycling.

Glass: Glass is primarily generated in the Food and Beverage Department. A small amount of glass may be generated from service bars in guest rooms. Most of the glass comes from beer, wine, liquor and food containers. Some processors require that glass be sorted by color-clear, brown and green. Processors with MRF's accept glass that is commingled. The glass is then color sorted by the processor.

Metal: Metal, aluminum and steel/bimetal, materials for recycling are generated through out the hotel/motel. Soft drink and juice machines contain aluminum cans. Steel/bimetal cans are generated in kitchens and food preparation areas. Equipment made from metal is located in various areas of the property.

Plastic: Plastic is found in the Food and Beverage and Housekeeping Departments. Guests also leave plastic bottles in guest rooms. The Society for Plastics Industry (SPI) developed a standard identification system that identifies different kinds of plastic resins. This helps plastic processors recover different kinds of plastic materials. The coding system has a recycling triangle with number l-7 which identifies the resin. Originally only plastic bottles were coded, however, now most plastic containers have the recycling symbol on the bottom. Check with your processor as to the type plastics they will accept.

Cooking Oil and Grease: Collect and strain cooking oil and grease from fryers and cooking in kitchens. Companies that recycle cooking oil and grease may provide collection containers for its storage and haul it from the hotel. This service is provided at no cost by some grease recyclers.

Furniture: Recycle used furniture by refinishing and reupholstering it. Furniture can be sold to other hotels, employees or donated to charitable causes. Refinish pool and patio furniture instead of purchasing new furniture. Aluminum patio furniture is recyclable with aluminum cans at most processors.

Equipment: Repair quality used equipment. Recycle old equipment with a metal recycling processor, trade it when purchasing new equipment. Give old equipment to vocational schools for feaching repair techniques to students. Keep it out of the hotel's waste stream.

Motor Oil: Establish a motor oil recycling center in the maintenance department for used oil from equipment motors. Oil collection centers for public use are located throughout the state. Ask your oil supplier about setting up an oil recycling program at the hotel for use by employees and vendors.

Construction Debris: Include collection and disposal of construction materials in contracts. Dispose of construction materials separately from the hotel's regular waste. Construction debris such as windows, doors and lumber are recyclable. Inform contractors of your recycling program and request that they comply while working on the property.

Landscape Waste: Include landscape waste removal in lawn maintenance contracts. If lawns and shrubs are maintained by the hotel establish a landscape recycling program. Shred fronds and clippings to make plant bedding material to save water, moderate soil temperatures, discourage weeds and reduce competition from grass. Call your local County Cooperative Extension Service for information about landscape recycling.

7

Service Development

Fundamental changes have occurred in the structure of most countries' economies since the Second World War, with services becoming the major sector of economic activity. The USA was the first service-oriented nation. At the beginning of the 1900s, fewer than 30% of workers in the United States were employed in services, compared to 40% in agriculture. Already by 1929, however, 55% of the working population was employed in services, and approximately 54% of GNP was generated by services.

Service employment in the US rose to over 66% by 1980; the trend in Europe was slower, though by 1977 52% of employment in the EEC was in the tertiary sector. In terms of the creation of new employment, since 1969 services have accounted for 95% of all new jobs in the US, and over 75% of GNP. Table 1 presents data on value added by services to GDP in a selection of relevant countries:

Table 1. Value added to GDP by services, selected countries 1987-1997

Country	1987	1997	Change (%)
Austria	64.1	68.2	4.1
Czech Republic	50.5	58.4	7.9
France	66.9	71.5	4.6

Greece	61.1	67.9	6.8
Japan	56.8	60.2	3.4
Spain	59.3	70.9	11.6
Switzerland	60.8	63.5	2.7
United States	68.3	71.4	3.1

This table gives a selection of countries to permit meaningful comparisons. These show considerable variations. Austria and Switzerland were included so as to compare two nations approximately equal in terms of geographical location and size of population. It may be seen that Austria is approximately 5% ahead of Switzerland in service value-added, perhaps owing to its lack of transnational corporations in manufacturing, such as Nestlé, Philip Morris and ABB, which are located in Switzerland. Also noteworthy is the share of services in the southern European states such as Greece and Spain, which have recorded considerable increases during the period under review.

The United States is included as the first service-based nation, with over 70% of value added in the service sector, which is now the norm for several European countries, including France and Spain. Although still low by EU standards, there has also been steady growth in services in the Czech Republic, typical of development in the former centrally planned economies. In contrast, growth has been much slower in Japan, mainly owing to the particular business environment, which is seen to act as a barrier to new entrants and to risk-taking.

The importance of service development is particularly marked in OECD countries generally, where the service sector accounts for more than 70% of economic activity, compared with less than 20% of activity in manufacturing. It is likely that the share of services will accelerate in the short and medium terms, given the ongoing shift towards knowledge-based, service-oriented activities in many countries. The most rapidly growing sectors are finance, insurance and real estate and business services; in 1995, for 19 OECD countries these sectors, taken together, exceeded US$ 1.1 trillion. By 1997 ten of the largest

20 corporations in the world by revenues were service enterprises. Service companies were also important by market capitalisation, with Microsoft placed first in the table at USD 407 billion, and Wal-Mart and AT&T sixth and seventh.

Changes have occurred on both the supply and demand side for services. Often it is not the type of human need that has changed, but rather the way of satisfying that need. Developed countries have moved from industrial, mechanical, nationally-focused economies producing mass consumer durables under a regulatory framework and in stable conditions, to globalised, de-regulated, client-focused economies characterized by rapid, discontinuous change, which is impossible to forecast in the long term. This has resulted in the development of "service economies".

The key difference between an "industrial society," which attributed value in terms of material objects that could be exchanged, to a "service economy" concerned more with the benefits derived from the performance and utilization of products (which may or may not be material objects) is that services have become the "indispensable production tools necessary to satisfy basic needs and increase the wealth of nations". Services are the absolutely central concept and it is the quality of performance rather than the quality of output that stands as the main criterion of evaluation: the critical factor is the quality of the customer's experience, the "outcome" rather than the "output". It is clear that these fundamental changes have impacted on the extraordinarily rapid development of service industries during the past 25 years in both developed and developing countries, and in many cases have provoked it. The international hotel sector is one such industry.

Classifications of Service Industries

Despite the importance of services in terms of their respective contributions to national economies, "[the] service sector is one of the least understood portions of our global economy". The diversity of services is also often neither appreciated nor fully understood. Historically, difficulties over definitions and classifications have posed problems for measurement of profitability and productivity. In the absence of a unanimously accepted classification scheme, many academics have developed different methods of classifying services. Over 16 different schemes were identified by Payne, using a wide variety of factors ranging from the type of service to the degree of labour intensity it required.

While traditionally manufacturing and services were in the past considered to be very distinct, the two sectors are now becoming more interrelated and the differences are blurring. One main reason is the role of information technology in overcoming the storability and the intangibility of services. The role of electronic commerce has been fundamental in this development, and services are becoming increasingly linked to products, if not inseparable from them. Outsourcing has also been a key factor in the development, as companies focus on their core competencies and leave more specialised companies to take care of peripheral tasks.

Earlier academic writing on services concentrated on the differences between goods and services. Four generic differences were defined, including intangibility, heterogeneity (variability) perishability, and simultaneity of production and consumption. Theses concepts are no longer universally applicable to all services, however, and it may be more useful to focus on the commonalities of services, looking at different ways in which service-based companies differ from those which are goods-based. According to Lovelock, there are eight generic differences between goods and services that are important for marketing, and the eventual potential globalisation of services. Table 2 shows the main differences:

Table 2. Key differences between products and services

Peculiarity of Services
Greater involvement of the customer in the production process
People as part of the product
Greater difficulties in maintaining quality control standards

Greater difficulty for the customer to evaluate
Absence of inventories
Structure and nature of distribution channels
Relative importance of the time factor

There differences are discussed in more detail in Lovelock, Yip and George, who state that the potential for globalisation of a service industry is dependent not only on these factors, but also on whether the service is classified as "people processing", as "possession processing," or as "information-based". Additionally, service systems for production and delivery can be categorised into "front office" or "back office (for hotel services, in a literal sense). "People processing" services by their nature involve a high degree of contact with front-line service employees and with the physical facilities. Many "back office" operations, however, can potentially be conducted at a distance, and with "possession processing" and information-based services there are many possibilities of reducing the physical contact with the client.

Increasingly, consumers purchase an intertwined package of goods and services. This has important implications for economies of scale and scope, and for the relationship between suppliers and consumers. The Internet has radically changed the method of transactions in such areas as education, banking, retail and financial services in ways which were unthought of just a few years ago; for example, it is no longer even necessary for a patient to be physically present alongside the consultant to have a medical diagnosis.

For the tourist and hospitality industries, information and communication technologies are changing the methods by which clients view, decide, select, and reserve accommodation, although the core service still, of course has to be provided when the client physically travels to the location and the unit.

The OECD uses the International Standard Classifications System (ISIC) that has four classifications of services. Hotels are found under category 1, with wholesale and retail trade and restaurants. However, while there are certain similarities between the different trades, hotels and supermarkets, for example, do not have much in common, and grouping them together makes many statistics unusable.

Specific statistics on hotels are provided by the World Tourism Organisation (WTO), which has developed a definition for hotels that in some degree does facilitate comparisons. To adopt a pragmatic approach, which will allow for measurement and comparisons, and services have been defined as follows:

> a diverse group of economic activities not directly associated with the manufacture of goods, mining or agriculture. They typically involve the provision of human value in the form of labour, advice, managerial skill, entertainment, training, intermediate services and the like.

CHANGES IN SUPPLY AND DEMAND FOR SERVICES

In parallel with the large increase in the role of service industries, there have also been major changes in the types of services demanded and offered. Many authors have analysed the underlying reasons for the increased and differing demand for services including Porter, Lovelock and Segal Horn. On the supply side, changes have resulted from such major factors as deregulation and globalisation; in demand, from changing family structures and increased leisure time. Table 3 lists the major factors that have altered the nature of services in the past two decades.

Table 3. Major factors impacting on supply and demand in services

Supply	Demand
Computerization and technological innovation	Greater affluence
Increasing internationalisation	Socio-economic changes including dual-career families; pressures on personal time; and fewer joint family activities, such as meals.
Increasing concentration,	Demographic changes including

domination of few MNEs	increasing numbers of children and older people who consume services.
Privatisation	Technological changes that have upgraded service quality or made entirely new services feasible (such as in medical care, cable television, and on-line data bases for personal computers)
Growth of Franchising	Urbanization, creating new services (e.g. security services)
Changing patterns of government regulation	More leisure time
Relaxation of professional association standards	Cultural homogenization
Increasingly complex management	Increasing sophistication leading to more and broader service requirements (e.g. in personal financial services)
Manufacturers as service providers	
Expansion of leasing and rental business	

The above table shows how many fundamental factors have had a direct impact upon the number and nature of services. Evidently, many such changes have taken place in developed nations. Changes have also occurred in the supply side, and again, while these are particularly in evidence in the developed world, there are signs that in the developing world (e.g. the case of India and the supply of computer software) changes in services will also have a profound effect on economies.

Even more fundamentally, Vandermerwe comments on the changing nature of customer demands and buying habits, showing that such needs have now changed profoundly. Fundamental reorganization is required within companies in order to move from industrial hierarchies to service networks, and so address the following axioms:.

1. Customers want solutions rather than core products or services;
2. Customers want more value for users;
3. Customers need to have access to suppliers' know-how and information throughout their use of the offering.
4. Customers want integrated global delivery service.

For companies to accept these propositions will require a paradigm shift in their interpretation of their operations, and in many cases of their raison d'etre. This is especially the case in the international hotel industry, where many hotel functions (such as the front desk) are often there for historical reasons; and often provide a negative value (for example through queuing periods) for guests. Many corporations entering into strategic alliances so as to provide clients with a full range of seamless service offerings (e.g. airline, hotel, car-hire service), deliverable throughout the world.

TOURISM AS A MAJOR SERVICE "INDUSTRY"

Although each industry sector is distinctive, it is useful at this stage to describe the specificities of tourism.

General industry characteristics

One obvious peculiarity of tourism is that the consumer comes to the supplier of the service rather than the reverse. Although tourism is often referred to as "the world's largest (and often fastest-growing) industry" it is not one industry, rather is it composed of a number of related and inter-related service sectors that do not fit neatly into the standard criteria for national accounts.

The Standard International Classification of Tourism-Related Services (SITCA) of the World Tourism Organisation

(WTO) identified 70 specific activities related to supplying tourism services, and an additional 70 activities at least partially concerned with them. An important consequence of this dispersion is that it is extremely difficult accurately to measure the full economic impact. Nevertheless, when looking at the sum of the parts, i.e. accommodation, meals, local transport, entertainment and shopping, statistics relating to tourism are of significant scale.

Tourism is taken here to be international tourism; this is estimated at only 10% of total tourism flows. Even so, by 1995 tourism accounted for over one third of the value of total worldwide services trade. In terms of output, travel and tourism are forecast globally to generate 11% of GDP. Tourism is a highly labour intensive industry, and is a major source of both employment and foreign currency, especially in remote and rural regions. Tourism demand (both international and domestic) is directly related to income levels; therefore the sector has grown as a function of increasing prosperity throughout the world.

In contrast to fast-moving consumer goods (FMCG) marketing, tourism marketing may be seen to consist of "slow marketing", as tourism projects require substantial investment in infrastructure and other facilities. This can run into billions of US dollars and take over a decade for results to be seen as is the case in certain countries in Eastern Central Europe.

One of the most important factors for the development of tourism was the invention and adoption of wide-bodied, large-capacity aircraft that facilitated low-cost airline travel. Tourism and many of its constituent parts (airline seats, hotel rooms, restaurant covers) is highly perishable and inventory cannot be stored. This accounts for the extensive usage of yield management techniques in the industry.

Tourism depends on various transport sources to deliver clients (with airlines accounting for 90% of international travel). Regulations concerning the ease of obtaining immigration entry and exit have a direct influence on the sector.

It is an important employer in developing nations. This may be attributed to the high growth rates of the tourism sector in relation to the domestic economy as a whole: already by the late 1970s there was substantial expansion by international hotel companies in developing countries.

Initially tourism was heavily concentrated in the developed world (focused strongly on Europe and the United States, which had over 70 % of market share). The share of developing countries now accounts for 1/3 rd of the total. Less developed countries have enjoyed strong increases (albeit from very low bases). Other factors that will have a bearing on the development in these countries are perceived political and economic stability, along with levels of transport and infrastructure.

Defining tourism services

Although it is a set of complex and interrelated service activities, we may usefully define tourism as "the activities of persons travelling to and staying in places outside their normal environment for period for not more than one consecutive year for leisure, business and other purposes". The tourism "industry" may therefore be considered from both the demand and supply side as those organizations and individuals involved in the consumption and supply of tourism activities.

Much of the quantitative information is generated from the demand side, focusing upon the numbers of tourists and sales generated. It is important to differentiate between an "international arrival", someone who crosses an international boundary, and a "tourist" someone who stays for at least one night in a foreign country. This distinction is especially important for this study, as the number of international arrivals in Eastern Central Europe has recorded spectacular increases, but for the hotel industry, it is the person who stays overnight that is the most interesting prospect.

These statistics may then be aggregated into the different industry sectors to provide a fuller picture of the contribution that tourism makes, both directly and indirectly, to the various industry classifications. Coefficients are then calculated to obtain the numbers of people employed in the different branches. Recently, real progress has been made in quantifying the full impact of tourism across differing industry sectors. After much ambiguity and inexactitude, serious attempts have been made to determine the contribution of tourism and the value added from the demand side.

Over the past fifteen years the Statistical Committee of the OECD has been working with the World Tourism Organisation and the UN to develop a set of Tourism Satellite Accounts. The creation of the TSAs is an important step forward, enabling more accurate comparability and credibility in data analysis of the direct impact, (e.g. in hospitality and tourism enterprises), and in indirect supplier sectors (e.g. in aircraft and ship manufacturing.)

Growth in international tourism

Just over half a century ago, in 1945, only 25 million persons crossed international borders. This number had increased by a factor of 17 to 425 million by 1990, and had climbed to 699 million ten years later. By the year 2020, it is anticipated that there will be 1.6 billion international tourists (a 64-fold increase) spending US$ 2 trillion. The World Travel and Tourism Council forecast that the travel and tourism industry would contribute 2.9% to Gross Domestic Product (GDP) in 2001 amounting to US$20.5 billion, which is anticipated to almost double by 2011 in dollar terms to US$42.5 billion. The travel and tourism industry employs over 4.5 million people, (3.2% of total employment in 2001), which is forecast to rise to 3.6%, or 5,824,000, in the same period.

As may be seen, there have been significant changes from the development of mass tourism from the 1950s onwards. Its geographical concentration, which saw 97% of tourism in 15 leading countries, all in Europe and North America, has been reduced. The industry has become more diverse and more "democratised". There have been other underlying trends, such as attempts to lessen the impact of seasonality (i.e. the preponderance of midsummer, north-south travel) through the staggering of school and work holidays.

A greater variety also appears through an evolution in the reasons for travel, which tends to blur the distinction between business and leisure. There is an overall trend towards shorter individual stays (but with more frequent breaks), and an increasingly varied supply of accommodation. The Meetings, Incentive, Congress and Events (MICE) segments have been a major driving force behind the development of business tourism. There has also been a major increase in different types of vacation offered, including cultural, sports, adventure, rural and cruise tourism segments.

Although the largest segment numerically and economically is still "sea and sand" tourism, there have also been significant attitudinal changes, resulting in a demand for more information on the local surroundings and for more activities to be made available.

From the original 25 million tourists who in 1945 visited just 15 countries, by 1997 over 70 countries had more than one million tourists, with the share of the top 15 countries slipping by 35%, from 97% to 62%. Also important has been the appearance of newcomers on the list from Asia, especially China, which by 1999 reached 5th position. Some countries of Eastern Central Europe have had fluctuating fortunes; Hungary, for example, reached 11th position by 1970, climbing still further to 5th place immediately after the fall of the Berlin wall, but then dropping back to 14th position by 1999. The Czech Republic (as Czechoslovakia) had also reached the top 15 destinations by 1970, but by 1990 had disappeared from view, only to return to 13th position nine years later. The most dramatic increase in international arrivals, however, has been in Poland. In 1982, the country was described as "a minor tourism destination". By 1999, however, Poland was within the top ten destinations worldwide, and attracted over 18.6 million international arrivals

There are specific changes that have impacted on the tourism sector directly. Some of the major factors are:

1. Increasing price and quality competition.
2. Tourist consumers who are increasingly becoming "dollar rich, time starved."
3. Polarisation of the market into either standardized low value-added tourism products/services or highly individualized/higher value-added services.
4. An increasing awareness of the need for sustainable development in tourism planning.
5. Continually increasing demand from consumers for more choice and activities, and improved interaction between service suppliers and consumers.
6. Increasing awareness of safety and security in consumers' perception.
7. Expansion of business tourism, including conventions and incentive tours
8. A general tendency towards more frequent, but shorter-duration, travel (improving seasonality), and interspersed with infrequent long haul, longer duration vacations.

Source: (Weiermair, 1995).

These trends have had an evident effect upon tourism supply. One interesting analysis of this effect is that of Poon who has applied the product life cycle (PLC) to the tourism industry, with the early stages of "mass tourism" appearing on the scene from the early 1960s. By the end of the 1980s, however, there were signs that this form of tourism had matured and the main characteristics of mass production - i.e. charter flights filled with sun-seeking travellers who passively accepted whatever was given to them - was unappealing to many new segments of the market.

Instead, the so-called "new tourism" was being offered as the alternative, based on sustainable development with diagonal integration and a combination of new technology and differently segmented markets. "New tourists", it is postulated, are fundamentally different from old tourists in a number of key areas, including higher income, travel experience and education levels, and greater environmental and cultural sensitivity. The result may be seen in the development of such tourism products as adventure holidays, with their mix of nature, fitness and excitement.

Economic Importance of tourism

The industry's direct services to tourists represent between 3-4% of GDP worldwide. Strong regional variations exist, however, and in some countries this may be as high as 10%. Travel and tourism is growing at about 3% per annum, again with differences according to region. In recent years, growth in Europe has been lower than average at 2.3%, but this masks intra-European differences, with Eastern and Central Europe recording growth of 5.2%. Again, regional variations range from 1.2% in the Americas to 2% in the Caribbean and Europe.

If indirect services are also included (i.e. those that provide infrastructure and supplies to the tourism industry), the tourism-related economy has been calculated to produce up to 11% of GDP, employing 8% of the labour force worldwide. As an approximate calculation, one direct job in the tourism industry induces one-and-a-half jobs in the wider, tourism-related economy. It is important to note that tourism includes travel undertaken for business purposes, not just for leisure.

The International hotel industry

The hotel industry is an important sub-sector of the tourism industry and provides accommodation and related services to tourists.

There are two main definitional problems in the industry that causes problems in the collation and comparison of statistics. Firstly there is the issue of what actually constitutes "a hotel". For example one may speak of hotels, private hotels, residential hotels, guest houses, boarding houses, lodging houses, bed and breakfast, inns, pensions, motels, auberges, posadas, et cetera, which in different countries may or may not fall within the remit of an hotel per se. In Europe, the problem is compounded, due to the peculiarly diverse range of provision.

The nearest to a universally accepted definition is that of the World Tourism Organization (WTO) which includes hotels within the classification of tourism accommodation as "collective tourist establishments", comprising: hotels, apartment-hotels, motels, roadside inns, beach hotels, residential clubs and similar establishments providing hotel services including more than bed-making and cleaning of rooms". The WTO specifies that in order to qualify as a hotel certain criteria must be included, including a minimum number of rooms (but the actual number of rooms is not mandated) and that the premises should be under a common management, with certain services, including room service, daily bed-making and cleaning of sanitary facilities being provided.

Although this is far from a perfect definition and is still open to individual interpretation by different countries, nevertheless the WTO states that, on a national basis, the total number of accommodation establishments covered in national statistics should represent at least 95 % of the total overnight stays in hotels and similar establishments. Despite the differences this means that there should, therefore be some consistency in reporting between countries.

The second problem in definition for the purpose of statistics is the hotel industry is composed of many different elements that are not always clear or easy to analyse. The hotel industry is notoriously difficult to "tie down", with some authors claiming that there may be several distinct industries operating within the overall "industry" structure. Three types of companies may be identified:

(1) Companies involved in constructing, developing and owning hotel buildings (e.g. Accor);

(2) Companies involved in managing hotels, with or without equity in the hotels under management, (e.g. Hilton International);

(3) Franchising companies, which develop hotel chains without being involved in either owning hotel buildings or managing hotels such as Choice hotels or Cendant.

A further complication, characteristic of the hotel sector, is that of "invisible" hotel companies existing with little visibility to the general public; although they may be of significant size, they exist solely through the management of hotels under different brand names.

Development of Catering Operations

Catering operations, as either a stand-alone facility or as part of a larger hospitality-related business, exist in a wide variety of formats, or styles. Most common are those that are readily identifiable as private rooms in restaurant operations, hotel facilities, and independent catering facilities. The increased demand by the international public for private function space outside of their own homes and businesses has led the catering segment of the foodservice sector of the hospitality industry to be a leader in the continued growth of both facilities and revenue. Forward thinking foodservice businesses from fine dining restaurants to delicatessens are incorporating catering services into their operations in recognition of the expanding market for preprepared foodservices. Offpremise catering and take-out services offer an excellent avenue for increasing revenue with minimal costs.

Catering services have been incorporated into foodservice operational styles, providing operators with techniques and methods for expanding the profit making potential of their businesses. The categories of foodservice operations offering:

1. Full-service restaurants
2. Hotel food and beverage facilities
3. Catering halls
4. Independent caterers
5. Private clubs

6. Contract feeding
7. Charcuteries and delicatessens

"Services" in the context of a foodservice operation refer to the opportunities management makes available to the customer to purchase food, beverage, entertainment, and ancillary services. Services include:

1. Table foodservices
2. Packaged take-out foodservices
3. Beverage services
4. Entertainment services
5. Business meeting services
6. Conference and convention services
7. Contract feeding services
8. Off-premise foodservices
9. Home-replacement foodservices

FULL-SERVICE RESTAURANTS

Full-service restaurants have the opportunity to offer a variety of catering services to their customers. Before any decision is made to offer these services, however, six important factors should be considered:

1. Location
2. Customer profile
3. Restaurant style or concept
4. Staffing capabilities
5. Restaurant physical layout
6. Cuisine and menu offerings

Location

The proximity of the restaurant to office complexes and centralized business areas will help to establish whether its catering service will be focused on business or social marketing efforts. Businesses in the twenty-first century are spread from urban centers to suburban locations in office parks. Central urban locations offer a concentrated market for both office delivery and take-out. A significant factor in developing the market for business catering is that service is generally required during the business week, leaving weekend periods free to service social business. In addition, locations such as museums, concert halls, and historical sites offer interesting venues to catered functions for both local businesses and conventions.Both urban and suburban restaurants can successfully develop social catering business. Suburban locations are generally more appropriate for social catering to private homes, clubs, churches, and other facilities. Delivery to urban locations can pose security and logistical problems, creating additional costs for transportation and service labor. Population density also affects the volume of anticipated catering business. Restaurants situated in rural areas with low population density cannot expect immediate high volumes of catering business. Areas of high-density population yield a variety of catering opportunities that steadily increase in volume through referrals and reputation. The location of the physical restaurant building plays a role in the type of catering services to be offered. Storage facilities, expansion possibilities, and access to major transportation routes are factors important to catering service production.

Customer Profile

Restaurants have the advantage of a built-in customer pool to whom they can market in-house and off-premise catering services. In addition, the attraction of being associated with a restaurant's reputation will help to expand the possible market to include new business and social clients. The market profile should classify customers as business or social catering clients, designated by income bracket. In addition the range of activities for which each customer pool will need catering services should be researched as thoroughly as possible. This will help in developing package programs along with potential menu programs and accompanying pricing concepts.

Style or Concept

The style, concept, and/or theme of the restaurant should be taken into consideration when planning potential catering services. Off-premise catering services do not necessarily have to blend with the facilities offered by the restaurant. On-premise catering services should, however, be designed to function within the restaurant facilities.

Facilities

Restaurants facilities are a major factor in providing on-premise catering. The ratio of catering functions to restaurant services that can be handled at a given time is dependent on the size and flexibility of the physical plant. Small private parties are often incorporated into the general dining room setting. Large parties must, however, be given facilities that are separated from the general public. This restaurant has a private dining room and bar facility with a dance floor that can also be used for additional restaurant seating during busy time periods and holidays. Flexible facilities such as these allow a restaurant to maximize revenues. Many restaurants that offer in-house catering schedule large parties, such as weddings, anniversaries, luncheons, and dinners, on days and times when the restaurant is not otherwise open. Often catering business must be refused because sufficient on-premise facilities are not available. When management is continually turning away catering business, a decision will need to be made as to how the restaurant will balance is future development of catering versus full-service dining services. Kitchen facilities play a major part in determining when and how catering service demands can be met. Kitchen equipment must be flexible, allowing for volume production to take place simultaneously with à la carte restaurant service. The kitchen cooking load and holding capacity of ovens and auxiliary equipment is important to determine if a kitchen is to be used to its full capacity during busy times.

Storage and refrigeration facilities determine the amount of food products available at any given time. The cost of waste from food spoilage due to lack of refrigeration and freezer space could dilute the profit from additional catering business.

Cuisine and Menu

The primary cuisine and menu offerings of a restaurant constitute one of the most important considerations for on-premise catering. As discussed earlier, off-premise catering services do not necessarily need to be the same as those offered at the restaurant for full-service dining. Purchasing and production requirements are crucial to the successful development of catering services. On-premise catering should offer menu items that duplicate the established menu as closely as possible in order to enhance production capabilities. Surrounding items, such as vegetables and starches, are the most effective area of the menu to duplicate. Kitchen production is more efficient when the number of surrounding menu items is limited. Profitable and effective purchasing for catering functions requires that the ingredients for menu items be the same or similar to those on the restaurant menu. This allows the purchasing agent to place orders for maximum volume pricing and quality.

Off-Premise Catering

The decision to participate in off-premise catering requires a thorough review of a restaurant's resources. A successful in-house catering program can lead management into perceiving that taking on the challenge of off-premise catering will not cause operational difficulties. Several problems must be anticipated regarding the off-premise facilities in which the final production and service for functions takes place. During in-house catering functions, service staff can often work between private parties and the dining room, filling in where needed, while the kitchen staff functions as usual, having planned ahead to handle the increased production load. In order to service off-premise functions, however, a separate waitstaff needs to be scheduled. In addition, at least one member of the kitchen production staff will be assigned to the function and therefore will be unavailable to the restaurant kitchen for the time period involved for the off-premise function.

Equipment as well as food will need to be transported both to and from the off-premise function location. Hot food and cold food must be kept at specific temperatures to prevent food spoilage. A refrigerator truck may be necessary at certain times of the year. Transportation routes and problems involving traffic congestion need to be considered in the timing of deliveries. Subcontractors may be necessary for equipment such as tables, chairs, dishes, glassware, linens, dance floors, and tents. The extent to which a restaurant commits to a full-service catering business determines the amount of extra effort required to handle the increased business. Planning for offpremise catering requires attention to detail. Contingency plans for bad weather, delays in food transportation, and the failure of on-site equipment to function properly should be established in advance so that problems can be dealt with expediently as they arise. Additional costs are incurred when solutions to these situations are not planned for in advance. Contract prices must include these contingencies or management will find that overall costs are higher than the fees charged to the customer for food and services.

Take-out service is the most profitable way for restaurants and catering services to increase revenues without increasing costs. Because take-out service does not require extra seating, it generally will not require an operation to expand its facilities beyond accommodating takeout customers with a pickup and waiting area. Kitchen production can plan and schedule to handle large increases in volume for short time periods. In addition, take-out does not require the additional expenses incurred by table service for glassware, linens, flowers, menus, and entertainment. Disposable dishes, plastic flatware, and a carrying container are the basic requirements for take-out service. Successful take-out service does, however, require planning to assure that customers receive menu items in satisfactory condition. As with the catering menu, the selection of take-out menu items should be drawn from the restaurant menu in order to maximize purchasing and production efforts.

Hotel Food and Beverage Facilities

The hotel food and beverage department provides food-related guest services throughout a hotel, conference, or resort property. Outlets for food and beverage services can include the following:

1. Full-service restaurant
2. Coffee shop
3. Catering facilities
4. Room service
5. Recreational areas
6. Lobby area bars
7. Food market or delicatessen

Food items for these combined foodservice areas are provided from a central kitchen, with the exception of large hotel facilities where satellite kitchens provide auxiliary production and service areas. An executive chef supervises production for all areas, consulting with the manager of each foodservice area. Of the seven food service areas, catering affords the possibility for the greatest profit, in addition to providing much-needed cash flow during periods when room sales are slow. Hotel catering services are usually classified as:

1. Business
2. Convention/conference
3. Social

The volume of catering service available to a hotel is based on the size and number of facilities that can be used for private functions as well as the availability of production and service staff and related equipment. The primary catering market for a hotel is based on three factors:

1. Location
2. Hotel facilities
3. Customer profile

Location

The location of a hotel determines the demand for in-house catering. Inhouse catering is defined as guest-related foodservice functions associated with business meetings, conferences, conventions, and social concerns. A hotel with a remote location focuses its catering efforts on the business booked into the hotel. Resort locations often provide additional opportunities for catering services at sports facilities and landscape venues throughout the property. A hotel with a suburban or downtown location can expect to develop a large volume of outside catering business, both business-related and social in addition to conferences and conventions.

The amount of combined catering business that can be handled at any one time in a hotel is based primarily on the number and size of the facilities available. The major function rooms are usually referred to as the ballroom areas. These are complemented by a series of meeting rooms, some adjacent to the main ballroom area and some located in other areas of the hotel. The main facility is a ballroom that seats 1620 guests for banquets and 2400 for theater style meetings. Adjacent to the ballroom is a series of connecting function rooms. These lead to another series of rooms that begin with an open reception foyer and lead to a second ballroom area seating 600 for dinner. An open terrace area is adjacent to these rooms, offering an outdoor reception, dining, and entertainment area. Private meeting rooms are available in other locations throughout the hotel.

Customer Profile

The customer profile for hotel catering services is made up of both business and social clientele. Customers who use the hotel as a location for a twoor three-day meeting consider meal functions as part of the total meeting package. This arrangement not only enables customers to work within a budget that covers the entire project, but also fits with an approach to planning catering functions that considers the success of the overall meeting rather than each individual function. The catering manager is generally given the responsibility of creating a menu plan for the entire program along with theme functions and special events. The target market for hotel catering functions is a customer who reflects the social, business, and economic profile of the community. Hotel catering services attract a wide base of customers with a broad range of requests. Customers who plan business functions on a regular basis are familiar with the catering staff and services of the hotel. Their needs are often quickly identified and provided for. Customers who plan social functions on a one-time basis, however, need more personal guidance throughout much of the planning stage. Pricing is a primary concern of this customer group.

Off-Premise Service

Hotel food and beverage departments are often requested to handle the arrangements for functions to be held away from the hotel property. If a company wants to retain as much revenue as possible from business, convention, and social customers, an off-premise catering policy should be developed.

Catering Halls

A catering hall is a facility dedicated to private parties with an on-site production kitchen and staff. These facilities can offer a wide range of both inhouse and off-premise catering services. The major factors that influence the market for a catering hall are:

1. Style or concept
2. Facilities
3. Customer profile

Catering halls offer customers a self-contained private function space independent of a hotel or restaurant. Many catering halls specialize in social functions such as weddings and plan the design of the landscaping and building to accent such functions. This design is also often aimed at a particular segment of the customer market, depending on location of the hall and the density and ethnic background of the surrounding population. Catering halls can also specialize in large functions, providing space for groups that cannot normally be accommodated in other facilities. Such halls are generally designed to host a wide range of social and business functions and a variety of special event themes.

Facilities

The flexibility, and often the originality, of function spaces are important to the success of a catering hall. Customers will travel considerable distances to social functions held in facilities that offer a unique setting or experience. Wedding receptions are often planned in locations over an hour away from the site of the ceremony so that the bride and groom can have a specific setting for their reception. Annual social functions are constantly looking for new or different themes to draw attendees. Catering halls can provide the setting for special event themes ranging from futuristic to country western, offering menu and entertainment packages to match.

Customer Profile

The customer target market for catering halls, like hotels, is based on the social, business, and economic profile of the community. Catering halls, also like hotels, are available in a variety of price venues, generally appealing to customers who are looking for menu and function packages within a specific range of prices. Identifying a customer group toward which to target marketing efforts includes surveying the competition, establishing a price range, and determining specific geographic areas of the community in which potential customers live based on demographic information such as annual income, family size, and age.

Catering halls have a great deal of flexibility regarding the type of business that they can solicit. The amount of off-premise catering that can be handled by an operation is limited only by the ability to service it. Location, transportation routes, and population density play a major role in determining this segment of the business. The ability to transport food and equipment and provide service and production staff are other limiting factors.

Independent Caterers

Independent caterers are private businesses offering catering services to the general public. These businesses operate with or without permanent facilities of their own in which to hold functions. Those caterers who operate without a formal facility must arrange for kitchen production and storage space. Kitchen space can be rented from schools, churches, senior citizen centers, and other institutional facilities with commercial refrigeration and production space. Storage space can be arranged for in self-storage warehouse rental facilities that combine short-term leases with flexible storage space and 24-hour access. Some catering companies have their own production facilities and enough warehouse space to handle large volumes of business. Chairs, floor covering, tables, and the table setup necessary for large functions are either sourced from a catering company's own warehouse or contracted through a local rental supplier. Many independent caterers focus on off-premise catering, maintaining only a small catering hall with limited facilities to house their production and storage needs. Their focus is on both business and social functions within a wide geographical area. In order to successfully cater to as many functions as possible simultaneously, a catering company has to be extremely well organized. Although equipment needs, staffing, and transportation are major concerns, the primary product, food, is the most important issue.

Menu Duplication

The key to successfully providing catering services to a number of functions simultaneously is the duplication of as many of the menu items as possible in order to minimize kitchen production. All surrounding items, salads, and desserts can be standardized. The main course selection should be limited to three or four items on any given day if possible. Menu items should be chosen for their ability to be precooked for completion in another location. Foods should have excellent holding properties and retain heat. This menu offers one appetizer, one salad and two choices of dessert. Dessert #2 is a chocolate dessert shell filled with frozen yogurt that can be preassembled or purchased frozen and transported to function locations within reasonable proximity to the main function kitchen. The gateau chocolate is contracted from an outside baking source, as are the breads. Whipped cream is served with the cake using a high-quality product from aerosol containers. The main course offers a choice of two meats, one poultry, and one fish. All four of these items fulfill the requirements for preproduction and holding properties. Two vegetables selections are offered, one appropriate for the meat dishes and one better-suited for the poultry and fish, but acceptable for any of the main course selections. Standardization procedures such as these minimize the problems of serving multiple parties in off-premise settings and ensure that food presentation is of the best possible quality.

PRIVATE CLUBS

Private clubs offer a self-contained facility that operates both full-service dining rooms and private function space along with a variety of food and beverage outlets. Private clubs are generally dependent on their membership for both dining room and catering business. Functioning as nonprofit organizations, clubs are, in many areas, prohibited by law from accepting or soliciting business from nonmembers. If such is the case private clubs may, however, cater functions that are sponsored by members and attended by nonmembers, allowing them to service both social and business activities. Factors that influence the success of restaurant and hotel catering services, such as location, facilities, and customer profile, also affect private clubs. Off-premise catering services can also be offered to private club members. Private clubs may also contract out their foodservice operation to a contract foodservice operator. An example of this type of arrangement is The Colonial Inn, Concord, Massachusetts, and the local country club. The management of The Colonial Inn turned to the local country club as a source of both kitchen production space and private function rooms. The Inn itself, a historic property located in the middle of a crowded town, has no available adjacent property on which to develop additional catering facilities. As a vehicle to increase their food and beverage revenues without incurring capital expenditures, the catering management team had identified a market for off-premise catering in the suburban Boston area adjacent to Concord. Without additional kitchen production space, however, the project could not go forward. In addition, during heavy business periods throughout the year, the Inn could not accommodate all of the requests that they received for private functions on property.

Through an arrangement with the local country club, management now not only has a satellite kitchen in a nearby facility but can also lease function space from the club on an as available/as needed basis. This alliance creates revenues and capitalizes on the needs of The Colonial Inn and the facilities of the club.

Location

Clubs are broken down into two general classifications: private clubs and country clubs. Private clubs are generally situated in the downtown areas of larger towns and cities. The club facility is used primarily as a location for meetings, private dining, social and business functions. Country clubs require a suburban location and usually promote a combination of golf, tennis, and/or boating as their primary recreational facilities. Private dining and catering facilities are located in the clubhouse. Acceptability of location is based primarily on general location and accessibility to major transportation routes. Downtown city locations can create problems regarding parking and security for people traveling into the city.

Facilities

Club facilities can greatly influence the volume of catering business. Clubs offering spectacular settings, access to garden areas for outdoor functions, or well designed interiors.

A club's ability to accept off-premise catering function business is controlled by the same laws and regulations as the solicitation of nonmember business. Beyond these restrictions, the same considerations for restaurants and hotels concerning transportation, staffing, production, and service apply. While the general specifications apply to almost everyone, each business will need to create specifications in each operational area according to its individual needs, availability of services, and budgets.

Contract Feeding

Contract feeding companies provide institutions such as hospitals and schools, as well as businesses, with in-house meal programs designed to meet specific needs. Food production and service is contracted for long-term periods with designated budget restrictions. The contract feeding segment of the foodservice industry has been steadily growing since the late 1980s. Major corporations such as Marriott International created large contract feeding companies to service a wide range of customers. In the late 1990s a separate company, Host Marriott, was created and now operates independently of Marriott International, serving institutional customers internationally. As food costs have increased and available labor decreased, institutions and businesses have turned to outside contractors to operate their foodservice programs. In most cases, these programs function as a service to either patients, clients, or employees; they are subsidized benefits, not profit making operations. By using a professional contract foodservice operator to handle this area of operations, businesses find that the quality of service increases significantly along with client and employee satisfaction. The major factors that can influence the success of catering services within a contract feeding location are:

1. Facilities
2. Customer profile

Contract feeding can be offered in a wide variety of dining room and catering facilities. The type and number of facilities that can be used for function spaces within a business or institutional setting will help to determine the volume of in-house catering. Catering functions in these settings are usually business related, but open area spaces can be used by private groups for social functions.

Customer Profile

Identifying the customer profile for large and diverse groups of people in a business situation is initially done by job classification level. Private dining room service and business meeting menu selections should be based on the expectations associated with the job levels of the people who attend these functions. The types of food products that will be popular in a cafeteria area depend on the profile of the employee population group. The proximity of the business to an urban center and the general education level of the customer group, along with local and regional food preferences and subsidy budgets, determine the range of food products offered.

Contract feeding companies working for a business or institution are often called upon to produce foodservice functions in off-premise locations.

CHARCUTERIES AND DELICATESSENS

This type of foodservice operation usually specializes in a regional or national cuisine. Menu items range from salads and sandwiches to fully prepared meals. The success of a charcuterie depends on two major factors: location and customer profile.

Charcuteries should be located where the customer profile supports the level of cuisine offered. Pricing for this type of foodservice operation is generally higher than that of delicatessens and supermarket preprepared and packaged food items; hence it requires a customer base willing to pay for gourmet food products. Smaller independent caterers often operate out of this type of retail outlet, using the facility for both production and storage in addition to a means of advertising and promotion. Delicatessens are food stores that provide a variety of food products and offer both eat-in and take-out food service. The location should be reasonably close to the targeted customer base. Menu item offerings range from sandwiches, salads, and pizza to preprepared meals-to-go. Take-out foodservice of preprepared foods is a major business segment for this type of foodservice operation.

The profile of customers for a charcuterie/delicatessen operation is drawn largely from the demographics of the population in the immediate area of the business. Primary customers can be identified as either employees of adjacent businesses or local area residents. A customer profile is determined for each of the two primary customer groups. For example, the twice a week business customer, with an average check of five to six dollars, could be in the mid-20s to 40s age group and low to middle income bracket. The local resident profile, with an average check of twelve dollars or more two times a week, may be in the mid-30s to 50s age group, middle to upper income bracket. Further investigation may show that the business employee is a morning coffee and lunch customer while the local resident is an evening customer for home-replacement meals.

Delivery of in-office and at-home preprepared food items can be a profitable extension of a delicatessen's business. From packaged lunches to party trays and platters, delicatessens often provide off-premise food items on a daily basis to a wide range of customers. This menu capitalizes on the restaurant's bagel bakery while also promoting salads, sandwiches, and desserts. Seating for these restaurants is generally limited due to their location in high rent districts. Marketing take-out and delivery services creates revenues without significantly increasing costs.

8

Managing Hospitality

Hi-Tech Evolution in Hotels

There are many interesting theories on the evolution of technology and the socioeconomic impact it has to nearly every industry. Hoteliers read and analyze opinions and case studies to help them make the appropriate technology decisions for their property, whether it involves installing a complicated Wi-Fi system, an innovative computer in the business center, or just a simple guestroom analog phone.

Sources such as the Wall Street Journal, Fortune Small Business, Mobile PC Magazine, Business 2.0 and many trade publications constantly introduce and analyze new innovative technologies. Hoteliers absorb the information and discover ways to apply new solutions to better serve the guest. Investing in hotel technology adds directly to the bottom line, achieving a potentially higher ROI. But, as with every other investment, you need to do your homework. Here are some technologies that have taken the hotel industry by storm with nothing but the guest's comfort in mind:

VoIP- Voice over Internet Protocol, according to the FCC, "is a technology that allows you to make telephone calls using a broadband Internet connection instead of a regular (or analog) phone line."

Now that telephone revenue in hotels is plunging due to a continuous increase in cell phone use, hotels must find alternate solutions for guestroom connectivity. With VoIP technology implemented in telephones, guests can make long distance and international calls anytime from their room at a very low cost. This only adds to the guest's comfort and overall experience at the hotel. For Hotel Commonwealth, one of the first hotels to use VoIP on property, it has been an exciting rotation of hi-tech trends that will yield substantial profit in the long-run on each call made using this technology.

Wi-Fi- Short for Wireless Fidelity, Wi-Fi is the most convenient way to access the internet. No more cable modems to keep guests cramped in their rooms to access email or search for data. In fact, according to Christina Valhouli's 4/29 article on Forbes.com, "The Best Wi-Fi Hotels," the number of properties installing Wi-Fi has increased from 2,500 in 2003 to over 6,000 so far in 2004. Adding to the fact that there are more than 12 million Wi-Fi users around the globe really makes this technology a must for any business hotel.

The problem that is constantly being addressed and monitored is how to avoid Wi-Fi piracy. Because the signal spreads, it is often times difficult to keep away intruders who park outside the hotel and use the hotel as a free hotspot. While piracy is an issue, more innovative software solutions promising a more secure network are constantly coming out, and they are worth investing in.

HSIA- High Speed Internet Access. Chances are if your hotel doesn't offer or has not considered offering HSIA in guestrooms, the guest has already taken several points off the overall experience. In fact, most hotels that appeal to business travelers offer this service complimentary, for convenience, and as an incentive for the guest to come back.

While still old school, connectivity with the traditional cable and modem, is still the mainstream way of accessing the internet from a guestroom, late at night, or early in the morning with the habitual continental breakfast. It is likely HSIA will be around for years to come, so having this in-room technology available to your guests is probably valued nearly the same as the amenity of cable TV.

Self-Operated Business Center- With new technology in hotels guests can operate the business center for more things than just checking e-mail, faxing or making color copies. Guests can now write reports, confidential memos in their room, hold the queue to print the documents in the business center and a guest password will print everything when ready.

Self-operated business centers are found mostly in smaller business hotels. They typically have one or two computers with full internet access that can be charged to the room, prepaid through internet cards, or charged to a credit card. There is also a main guest fax, copiers and color laser printers. Some hotels have in-room faxes as an additional business amenity. The Hotel Commonwealth was designed to meet the standards of the digital age, and therefore offers complimentary in-room Wi-Fi and HSIA instead, encouraging e-mail rather than fax or paper correspondence.

With so much new technology outdating its predecessors, it is a challenge to keep up. Answering machines, walkmans, and even cd players seem like ghosts from ancient times. And they are. The era of complete comfort and convenience is here to stay. And as long as people know how to maintain a balance between technology and natural life, then this advancement may be a healthy evolution.

On that note, we all know that to succeed in hospitality means to maintain and elaborate on the very basic elements from simply shaking hands and adding that personal touch, to going out of our way to make things happen, being constantly innovative, establishing good eye contact and sharing a smile with the guests. This industry was built on the essence of providing and fostering personal attention. Therefore, we cannot allow this philosophy to be lost in the new age of technology. Pursuing technological advancement is secure only when it enhances the personal attention offered to the guest, and not when it replaces it.

PROTECTION BY LEGAL ATTACKS

How can you dramatically reduce legal attacks while building a wall of defense around your company? Follow these ten simple steps and you will be on your way to protecting your company by building a fortress of defense.

Investigate and take action on all reports of harassment and discrimination

All managers and directors must report to the human resources director or other personnel any and all incidences or claims made by employees in connection with sexual harassment or any other type of harassment or discrimination. If managers do not report the incident or claim and an investigation does not occur, the company is at risk. Every employee should undergo annual anti-harassment and anti-discrimination training.

The company's policy of prohibiting harassment and discrimination of any kind should be clear, in writing, and effectively communicated to every employee. Newly hired employees, as well as seasoned veterans, need this training. During any investigation, if new incidences are discovered, those incidences should also be thoroughly investigated. A report, which includes written statements of witnesses and the resolution of the claims, needs to be well documented and placed in the appropriate personnel file.

Make sure all agreements with vendors and suppliers are in writing and their insurance has not expired

The financial officer of each hotel or restaurant should undertake a careful review of all agreements. Be particularly mindful of indemnification clauses (in which one person agrees to protect another person from claims by third parties), waivers and releases, automatic renewals of the term of the agreement and assignment clauses that prohibit the hotel or restaurant from transferring the agreement to a new purchaser or new operator. Make sure all vendor and supplier agreements allow for the hotel or restaurant to cancel for any reason or no reason on thirty-(30) day's written notice. Obtain current certificates of insurance from every vendor and supplier and ask for updated certificates periodically. Make certain that the insurance certificate lists the hotel or restaurant operating or managing company, as well as the owner of the establishment, as additional insured parties on the policy.

Keep accurate maintenance logs on all equipment

Many cases have been settled or lost in court due to the failure of the property owner to maintain accurate maintenance logs on equipment owned or serviced by a hotel. This includes the monthly, quarterly or annual maintenance checks on elevators, bicycles, fitness equipment, pool equipment, etc. It is important to maintain clear and accurate records to show a judge, jury or opposing lawyer the diligence that went in to ensuring the property was safe. Among other things, the records should reflect the times and dates of inspection, any and all repairs made and the name(s) of the qualified employee(s) responsible for maintaining the records. Accidents happen, of course, but an accurate maintenance log will go a long way in reducing your liability.

Follow all rules for proper notification to guests of in-room telephone charges.

Due to the popularity of cell phones and PDAs, hotel guestroom telephone usage has declined. However, many hoteliers may not be aware that there are federal and state laws stating what must be clearly communicated to your guests prior to them making telephone calls from guestrooms. Written notice of all fees and charges must be given to the callers in the guestrooms as directed by the statutes, and should be featured in a prominent location. The in-room guest directory should include the following information and applicable charges: local calls, directory assistance, alternate carrier access (for other long distance carriers than the one the hotel does business with), toll free calls, international calls and long distance calls.

Don't send unsolicited faxes to customers without their written permission

The 1991 US Federal Telephone Consumer Protection Act prohibits the sending of unsolicited fax advertisements in all states. Specifically, business owners are not allowed to send unsolicited faxes that advertise the commercial availability or quality of property, goods or services to customers without first obtaining their permission. The recipient of such faxes can successfully sue the sender and collect $500 per fax. Plus, if they can prove you willfully violated the law, they could recover $1500 per violation. Although some companies are challenging this law based on First Amendment issues, until the law is changed, compliance will eliminate these claims.

Have all employees who pour, serve or sell alcohol beverages, and all managers, attend an alcohol awareness program

Not only is it important for this training to occur, your legal adviser (and most likely, your insurance carrier) will insist that your employees learn the legal basics of pouring, serving and selling alcohol beverages to your patrons and guests. Most states have "dram shop laws" that require establishments to adhere to alcohol safety and awareness standards. If someone leaves your establishment and is personally injured or injures another person or property, the patron's lawyer will look to the establishment to see if alcohol was consumed on the premises. Be smart and implement preventative measures now.

Have a written email policy prohibiting jokes, off-color remarks, and other potentially offensive activities.

As part of their lawsuits against companies, plaintiffs are requesting that defendant companies produce all email transmissions in order to collection incriminating information about the company and its employees. These messages, although may appear to be deleted by each user, are not gone forever. They actually are maintained on back-up tapes that carry this information forever (at least a very long time until they are erased by an IT professional). Remind employees that although email messages appear to be a quick and casual way of communicating, it can be hazardous to your company's bottom line. Advise them not to put anything in an email message that they would not want to see as a headline in your daily newspaper.

Every company's written email policy should require all employees to avoid creating, viewing or forwarding offensive or inappropriate emails or Internet postings related to a person's race, color, national origin, religion, sex, sexual orientation, age, disability or veteran status. Also, they should avoid creating, viewing or forwarding jokes, cartoon, games or chain letters. The email policy should also provide for immediate disciplinary action or termination of employment, if necessary, for violations.

Be aware of scams

There are numerous scams ongoing for many years that managers need to be aware of and need to educate all employees. For example, con artists are targeting companies with the Nigerian letter scam. This scam begins with a letter or message that is mailed, faxed or emailed informing a business that there is a certain surplus of funds that exist and by providing your business bank account information to hold these funds for a short time, the business will receive a portion of these funds. This is an attempt to empty your company's bank account. Also, be aware that certain companies may be marketing a hotel directory that is not pre-approved or in line with your company's marketing strategies. An unsuspecting employee may obligate your company by agreeing to a long-term commitment to have your property listed in the directory. A good resource for learning more about scams can be found at www.snopes.com.

Inform your guests about laptop computer theft

In too many cases, laptop computers are left in hotel meeting rooms and are easy targets for thieves. To anticipate your guest's needs, provide a safe to store laptop computers when not in use or when the group is taking a break from their meeting. Provide computer-cabling locks to assist your guests in securing their laptops. Lock or control entry to meeting rooms when not in use. While not a preventative measure, hoteliers can also post a warning in the area so that the hotel is not responsible for stolen or damaged property. Prior to posting the sign, review your state's limitation of liability statute to determine your liability in such cases.

Don't release guest information to anyone other than the guest

Releasing guest information to the wrong person can result in a claim being filed against your property. Guests have a certain expectation of privacy that needs to be acknowledged and recognized by all employees. A written policy needs to be established whereby (i) the registered guest makes the request in person and either the guest is recognized by the employee or provides sufficient identification or (ii) the guest submits a signed, written request and the hotel or motel is able to verify that the signature belongs to the registered guest. Certain exceptions to the policy would include those circumstances when the hotel or motel is compelled by law to produce the guest information (a court order or subpoena, for example) or when a crime has been committed against the hotel or motel.

Depreciating Assets of Hotel Owners

Owners and investors in the hospitality industry should be watching carefully to see how changing tax incentives will affect their businesses. One should be aware of the more generous depreciation schedules associated with qualified leasehold improvements. In particular, this may be the time for investing in the latest computer software and for remodeling projects associated with the installation of hardware.

Many of the recent changes in the tax law were intended to provide incentives for small and large businesses to spend more on technology, machinery, and other equipment with the hope of creating US economic expansion. One method used was to increase the amount of property that could be expensed directly, rather than having the cost recovered through depreciation (the section 179 deduction). The IRS defines depreciation as an allowance for the wear and tear, deterioration, or obsolescence of the property.

The increase in the 179 deduction may be the most beneficial change for smaller businesses. It allows the expensing of part or all of the cost of qualifying property that is placed in service during the tax year. Qualified property is defined as new or used depreciable tangible personal property purchased for use in the active conduct of a trade or business. The Act expanded the definition of qualified property to include off-the-shelf computer software purchased during tax years 2003, 2004, or 2005. The maximum amount that may now be expensed is $100,000 and the spending limit begins at $400,000. An investment in qualified property exceeding that limit would reduce the deduction, dollar for dollar, until spending reached $500,000 where the entire benefit would be lost. Beginning with tax year 2004, the section 179 deduction and spending limits are indexed for inflation. It is important to note that a 179 deduction may not put a taxpayer in a net operating loss position.

While US Congress is often generous in its use of depreciation, the IRS does not always look at it the same way. The different classification of an asset's useful life often placed the IRS and the taxpayer in adversarial positions. To reduce the conflict, at the end of 2003 the IRS issued a field directive to its agents on the proper rate of depreciation in the restaurant industry. It provided a very detailed listing of assets and the recommended recovery period. Although the guidance is not precedent, the Service has instructed its examiners not to raise the issue if the guidelines are followed. It is important that a hotel review the list and compare its position with that of the

Service. Expect tax cuts and incentives to be widely discussed during the upcoming months. The economy is always on a politician's radar screen, but never more so than during an election year – and a Presidential one to boot.

Increasing Productivity and Profit Margins

It was great news for the International Development Research Council (IDRC) when their executive board decided to host two tradeshows this year instead of one. However, for Angel Fortenberry, show manager for IDRC, it meant double the work with the same budget and the same amount of time. Learn how online technology helped Fortenberry pull everything together and create two successful events. If you're like me, you've probably noticed that your to do list seems to be getting longer and longer every day. Meanwhile, there are still only 24 hours in a day and we still only have two hands to handle all of this extra work. It seems that the only option is to become more efficient in order to handle more work in less time. Nevertheless, we all know that's easier said than done. This was the situation for Angel Fortenberry, show manager for the International Development Research Council (IDRC), located in Norcross, Georgia. Her executive board decided to host two trade shows this year instead of one. This was great news for the association, with both shows generating an excellent net profit margin. However, this meant that Fortenberry faced the challenge of finding a way to get her job done more efficiently, and FAST! She turned to technology.

Fortenberry found the answer to her challenge in a web-based solution that enabled potential exhibitors, attendees, the show manager and administrative staff to access the event through a link on IDRC's web site. The solution includes an online booth sales module, registration system, invoicing, and confirmation system and more that would help save time and money. "Now," says Fortenberry, "exhibitors will never misplace their exhibitor prospectus or exhibitor kit again. And, I am never more than a click away from the information or reports that I need."

Here are five ways Fortenberry was able to save time and money with the exhibitor management portion of her online tradeshows.

Exhibitor Prospectus & Exhibitor Kits

Fortenberry saved thousands of dollars by not printing the exhibitor prospectus or exhibitor kits at all, rather putting them online. Since her two shows were technology focused, she decided that if a potential exhibitor did not have the ability to access the prospectus or kit online, that they would probably not be interested in exhibiting in her show anyway. If your exhibitors aren't all tech savvy, you can still download and print an attractive and economical copy from the website and fax or mail it to them.

Online Booth Sales

In the old days, potential exhibitors would call Fortenberry expressing the desire to purchase booth space. Then, Fortenberry would repeatedly fax marked-up floor plans, applications, confirmations and invoices, in order to execute dozens of space buys. By putting the process online, Fortenberry eliminated all the paper shuffling as well as the politics involved in assigning booth space. Exhibitors can see a real-time floor plan online, color-coded to show which booths are reserved, the company occupying each booth, and which booths remain open from which to choose. The entire selection process, from actual booth space selection, to making the request and making the payment online not only makes life easier for the show manager, but has earned rave reviews from exhibitors as well. Fortenberry said that her exhibitors have told her that the process as 'very simple' and 'a piece of cake'. She was able to sell 40 percent of her booth spaces in the first two days, all online as well as cross-market her two shows.

Invoicing

Fortenberry wanted to improve the time consuming process of

querying, setting up each invoice, printing and preparing each for mail or fax. Even with her established accounting system, she found herself spending a great deal of time preparing invoices. Now in a matter of clicks, Fortenberry can select a group of exhibitors, create and email the invoices, along with all the necessary forms the exhibitor would need to make the payment.

Confirmations

Mailing or faxing confirmations is a thing of the past for Fortenberry. For example, after exhibitors register online Fortenberry generates the confirmations and sends them electronically along with booth set-up instructions and a link to the online exhibitor kit, saving Fortenberry a significant amount of communications costs. "I also take this opportunity to get in a few reminders like the urgency of completing their online booth, so I can get their program copy downloaded from one of the great reports the program will generate," says Fortenberry. Furthermore, Fortenberry's online program tracks every confirmation sent, so she always has twenty-four hour access, from anywhere, to an up to date record of every confirmation sent.

Creating the Printed Program

No more re-keying exhibitor descriptions and contact information for her printed program. Fortenberry found that her technology solution was very helpful in saving her preparation and editing time for their printed program. She could access the all of the information keyed by exhibitors into their online booths, and download it to create accurate, timesaving files. Fortenberry explains, "I simply copy and paste the information from the program, paste it into a word document, edit and send it along to the marketing department. No more re-typing and proofing copy before it goes to press. The exhibitors supply the information and I just pass it along. Another time saver!"

High speed internet access in Hotels

Wireless or wired high speed Internet access ("HSIA") has become the new "must have" amenity for the business traveler in hotel public areas and guestrooms. Are your guests aware of the potential risks associated with using HSIA in their guestrooms or public areas? Are you responsible for advising them of these dangers? What do you do when customers complain about systems being down or not functioning properly? How can you lessen your exposure when a hotel employee's attempts to help a guest end up in disaster? These questions need to be thoroughly examined and answered prior to, or immediately after, installing a guestroom or public area wired or wireless Internet access system.

Currently, hospitality managers are not required by law to provide HSIA, whether wired or wireless, in hotel guestrooms or public areas. Hospitality managers, however, must provide an environment that is safe and secure for the patrons and guests while on their hotel or motel property. Once a hotel or motel provides a new service to its guests (such as providing HSIA) the law requires that they act prudently and use reasonable due care in providing and fulfilling the service. Generally speaking, the standard of care required by hospitality operators requires that a manager apply the same diligence as any other reasonable hospitality manager in a similar situation.

Standards of care are constantly evolving, however, reasonable care generally requires the manager to correct potentially harmful situations that they know exist or that may be reasonably foreseen to exist. For your guests' protection and your protection as a manager or owner of a lodging property, your guests need to know about the inherent risks in using HSIA in their guestrooms and in the public areas where wired or wireless service may be available. Once you begin providing HSIA services to your guests you are required to provide and perform the service in a safe and reasonable manner. Here are a few suggestions that are aimed at protecting the legal liability of your hotel or motel:

Make it clear to your guests that the hotel does not directly control the Internet provider company nor the system providing wired or wireless HSIA to the guestrooms or public areas. Your literature or brochure describing this new amenity should expressly state that another company, independent from the property manager or owner, is providing the service and not the hotel or motel. Of course, you need to find a very reliable source to provide the HSIA service to your guests; however, the guests need to be made aware that since you can't control the service, there is no guarantee.

It is critical that your Internet service provider or vendor goes through an extensive certification process whereby a complete security assessment is examined and security protocols are put in place prior to retaining them to install and provide the service. Don't use just any vendor to supply your Internet service. Do your homework and make sure that appropriate security measures are in place such as adequate firewalls and other features that are becoming standard in the industry for the protection of your guests and your hotel. You also want to make sure that the provider of the guestroom and public area HSIA system has security measures in place so that your own property management and hotel computer systems are not compromised in any way.

The Internet provider should include an on-line tutorial during the initial connection to the Internet whereby the guest must read and agree to the terms and conditions of use. The guest would be required to read through these paragraphs which will put the guest on notice of the inherent risks associated with using the Internet system provided. Providing notice of the risks is as simple as requiring them to electronically agree that they are aware of the significant security and privacy risks associated with accessing or transmitting information over the Internet, whether wired or wireless connections are used. Also include a clear disclaimer whereby the property and the Internet service provider are both protected stating that they are not liable for any loss of data, file corruption, hacking or any damage to one's computer while using the Internet service. If the property allows the guest to print documents to a central printing resource in the hotel's business center or at the front desk, the guest needs to know that security, as to the contents of the documents, is not guaranteed. The agreement must require that the guest accept the terms of the arrangement before logging on to the Internet. If an on-line disclaimer is not available, consider having your guests sign a written disclaimer prior to using the HSIA service. This disclaimer should contain the same items as described above. Make sure you keep the signed waiver on file in the event there is a claim brought by the guest in the future.

Your HSIA service needs to provide support for a VPN (Virtual Private Network) so that guests will be able to connect with their end-use home servers with confidence that security features are in place.

Ritz-Carlton has developed a concept called Technology Butler ? whereby trained employees are available 24/7 to provide technological assistance to guests on various computer and Internet access needs. Other hotels and motels have or will most likely follow this service trend since guests will have questions about wired and/or wireless services provided in the hotel or motel and how to access the Internet. It is very important that a written procedure be created setting forth the rules for employees to follow when approached with a guest's computer needs request.

Although not exhaustive of all procedures, such written procedures need to include, at a minimum:

— Untrained employees should never attempt to help a guest with his or her computer needs.

— Those employees who are trained to assist guests with technological needs should never personally touch the computer keyboard without the consent of the guest.

— The trained employee should keep the guest informed of every change to the computer settings or documents that will be made. Constant communication with the guest is essential.

— All employees assisting the guest with his/her computer needs must find out what the guest has already attempted to do up until the time that the trained employee begins to provide assistance.

— If the employee does not know how to solve the problem, he/she needs to call for assistance. No employee should attempt to do something on a guest's computer that they wouldn't do to their own computer. One simple error could be a very expensive one for your property.

— When possible, remind guests that wireless systems are not secure. Only non-critical email and basic Internet surfing should be accessed.

— Trained employees need to remind guests not to make any Internet purchases or do anything that could expose them to liability or potential identity theft, such as checking bank records or credit card statements, on wireless systems. The guest needs to know that a nearby hacker, who the property cannot control, may intercept his or her password and steal personal information of the guest.

— Be prepared to answer questions about whether the Internet sites visited by guests will be recorded somewhere on the hotel's computer system.

— Keep a detailed written log of all guest computer service requests including the date and time of a request, the guest's name, room number, specific request for assistance, action taken and the follow up with the guest.

As more hotels and motels begin to provide, and continue to provide, HSIA systems for their guests, technology will advance and ultimately enable HSIA systems to be more secure. Meanwhile, take these steps to place your property in a much better legal position in the event something goes wrong.

Building a Website

Building a website is a fantastic way to reach new clients for your business, to earn some additional income, or to generate enough income to allow you to leave the rat race. But if you're going to have a website you need to ensure that it's effective, it works and it produces results. The biggest single mistake made with a website is that, once it's up and running, there aren't sufficient people visiting it to produce the results you want. This is a mistake you can so easily avoid if you build your own.

If you don't have web design or programming skills you may think you can't do it yourself. I'm here to show you that you can. Building your own website is easier than you think and there are many advantages to doing it yourself. Firstly, you uderstand and know your business better than anyone else. You'll be in complete control of every aspect of your website and you won't have to wait for changes and updates to be made, so you can ensure that your website has all the essential elements to work successfully.

Most professional web designers may well produce a website for you that looks wonderful and could win awards. But a great looking website isn't so good if there aren't enough people visiting it and you're not producing sales and generating income. Your website needs to be effective and produce results. This is easy to achieve when you have the right tools and knowledge. You'll put the same amount of time and effort into a website that sits out there unknown as to one that works, so why not go for one that works?

You need to have an overall plan for your website; a plan that includes the structure and layout of your site, the look and feel of it, and what will make it work well in order to attract huge amounts of traffic and generate income. Also you must submit your website to the search engines, sending out an email newsletter and much more. However, building your own website doesn't mean you have to do all the work; you can delegate some of the admin tasks to someone else.

So, where do you start? To make your own website successful, you'll need to register a domain name, purchase hosting for your website, consider the design and layout and most important of all, promote it. Promotion of your website is an essential process and this means getting listed and then ranked highly on the search engines. Your site needs to be found. The search engines enable people to find you and without these visitors, your website is likely to end up the same way as most do and that's with little or no traffic - hence few or no sales. This is when you hear people say that the internet didn't work for them or their business. It's not that the internet doesn't work; they just didn't know how to make it work for them. The internet is truly a remarkable opportunity and you can tap into this now.

You need to know the purpose of your site. Is it for people to find out more information, to produce sales leads, or for passive income? Whatever the purpose, the single most important thing you need to do to make your website successful is to attract traffic to your site. That means people visiting it and if this is done correctly, this will be targeted visitors looking for information and ultimately the solution you provide. Hence, your site makes sales and you'll be streets ahead of your competition.

You may be wondering if building a website yourself is going to be too time-consuming. Yes, it is going to take some time, but it's surprising how much can be achieved in as little as 1 - 2 hours per week. It can be done more quickly than you may think.

Building your website won't happen overnight and it will take some effort. To make your own website truly successful, you'll want to include other elements such as a newsletter, forms and auto-responders. These all add to the effectiveness of a website and ultimately lead to sales.

Now you're probably thinking that building your own website is beyond you. You might be saying to yourself "I don't know enough about all this", "I don't have the skills or knowledge required" or "I wouldn't know where to start or how to plan what needs to be done". It's understandable that you're having these thoughts because there is quite a bit to consider in order to make your website successful. However, there's a very simple way to do it that doesn't require any technical knowledge.

So what are the ingredients needed? You and your passion or knowledge of your business or product plus some motivation. Now add the most powerful tool available and hey presto, you're well on your way to a successful website that triggers sales. This powerful tool is SiteBuildIt (SBI) and is an all-in-one site-building, traffic generating, and site-marketing product that makes it simple for you to build a website that produces results. It provides everything you need; all the information in a step-by-step guide, strategies and tools. You don't need search engine expertise, technical knowledge or have to hire an expensive consultant.

Five ways to Better Forecasts

While advanced mathematics and enormous computational power have improved significantly, few would argue that forecasting is an exact science. That's because at its core, forecasting is still mostly a human dynamic where accuracy is dependent upon:

— asking the right people the right questions;

— their willingness to answer truthfully and completely;

— the ability to separate the meaningful elements from the noise; and,

— the openness of the forecaster to suggestions of process improvement.

Consistently good forecasting isn't a mathematical exercise performed at regular intervals (e.g. quarterly) as much as it's an on-going process of gathering and evaluating dozens or hundreds of points of information into a decision framework. Then, when called upon (e.g. quarterly), this decision framework can output the best forward-looking view grounded in the insights of the contributors. While software can facilitate process structure by prompting for specific fields of information to be included, it cannot make judgments on the quality of the information being input. Garbage in; garbage out.

Which might help clarify what Yogi was referring to above. That even though baseball statisticians have access to over 100 years of data loaded into high-speed computers at their fingertips, the human element in what happens with the very next pitch makes it nearly impossible to forecast (with any acceptable accuracy) who will win the game, never mind the pennant.

Benefits of Better Forecasting

Marketing as a discipline benefits from better forecasting in at least two specific ways.

— First, sound forecasting processes form the foundation of an "early warning" system to alert the rest of the organization to the need to rethink its market orientation. In essence, forecasting becomes the rudder that can help your company stay the course, change directions, or navigate uncharted waters with confidence. As such, marketing migrates from being a tactical player to a strategic resource for the CEO when forecasts become more accurate, timely, and reliable.

— Second, better forecasting helps marketing respond to today's dashboard-equipped CEOs and CFO's who are continuously pressing for better ROI metrics and project accountability. Building confidence and credibility in forecasting helps marketers secure and allocate resources to new initiatives as the organization begins to trust that the forecasting prowess will provide a more reliable measure of future profits to be gained through today's initiatives. Five ways to improve the quality of your forecasts:

1. Be Specific.

As simple as this sounds, knowing exactly what you are forecasting is the most important step to success. It might seem pretty obvious that if you want to forecast sales, forecast sales. But what question are you really trying to answer? Unit sales? Gross margin? Market share? Customer value?

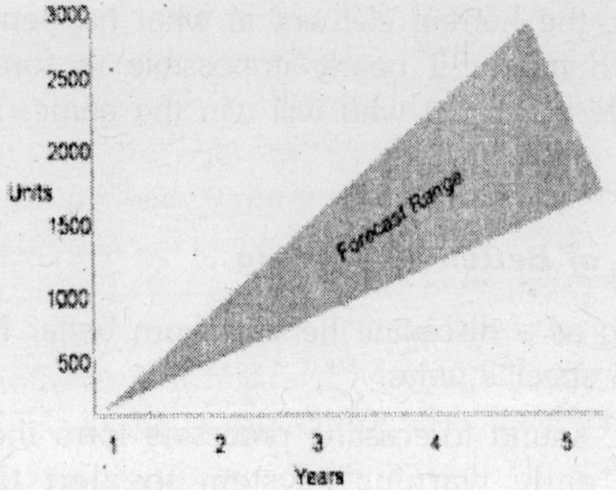

Also, what period of time do you need to cover? The longer out the forecast goes, the less reliable it is in the out years. This becomes especially important if your forecast is intended to anticipate the market size of a new category that will cost 10's of millions or more to enter.

In general, forecasts fall into one of two categories: operational and strategic. Operational forecasts manage the existing organization one or two steps ahead of today's reality. Strategic forecasts look further out into the future to help focus the company's long-range planning. In mature market categories, (toothpaste, personal computers, pet foods, etc.) the operational time horizon could be 2 to 5 years and the strategic 10 or more. But for younger or more turbulent categories (mp3 players, feature films, etc.) operational windows could be measured in months, while strategic horizons are maybe a year or two at best. In the early days of search engines, Excite was seeing traffic double every few weeks and new competitors entering daily.

Operational forecasting was on a horizon of 10 to 30 days and strategic was only 3 to 6 months since no one could realistically see any further ahead. It's tempting to try to divine the future and seek competitive advantage. But risk factors increase as certainty declines, and if your company doesn't have a "bet the farm" culture, it is likely a waste of time trying to look too far ahead.

Finally, what degree of accuracy do you need? BASES, ACNeilsen's famous database forecasting process for packaged goods products, has amassed the lessons of thousands of product launches in hundreds of categories across dozens of countries over a 25 year period. And yet their forecasting accuracy is still +/- 20%. Not to suggest that you might not do better in your own markets, but understanding the economic impact of being in a margin of error of 5% versus 50% will guide the decision about the forecasting process required and the appropriate amount of resources and effort necessary to achieve the goal. The CMO graveyard is full of those who missed forecasts by a wide margin due to faulty expectation-setting and/or misalignment of process to goal.

2. Be Structured

There are many reasons to take a structured, methodical approach to forecasting. First and most obviously is the importance of not leaving out key information which might affect the forecast. Also, there is the quality control factor and the benefit of double- and triple-checking all the assumptions and formulas. But among the less obvious benefits of structure are:

— The removal of personal biases which might unknowingly be causing participants to filter their inputs or interpretations;

— The continuity of consistently improving upon the process over time, regardless of turnover among key input or executional resources;

— The auditability of the approach to determine where things might have gone awry at various steps in the process; and,

— The confidence your rigor will inspire when others evaluate your work and are by necessity forced to accept some subjective judgments and assumptions.

Structure needn't be costly or time-consuming. In its simplest form it is taking the time to map-out and document all the inputs into the forecasting process; describing (in writing) the apparent relationships between causal factors; noting all assumptions and calculations in an easily referenced manner; and recording the accuracy of the resulting forecasts over time, alongside observations on emerging factors which might have influenced the results.

3. Be Quantitative – with or without "data"

If you have lots of historical data at hand, quantitative forecasting methods such as moving averages, time-series analysis, and exponential smoothing create a much greater likelihood of developing a strong forecast provided you have enough historical data to use them. But even if the only data you have are a series of "finger-in-the-air" estimates, you can still take a more disciplined quantitative approach by building simulations which explore the "what if" scenarios often hidden in best-guesses at average outcomes.

Regardless of the quantitative approach you use, keep in mind that like power tools, mathematics can be really dangerous in the hands of the inexperienced. Hiring someone with strong statistical skills to determine the most appropriate quantitative method(s), given your data (or lack thereof), provides yet another comparison point to check against your experiential judgment. Even if you choose to disregard the forecast derived by crunching the numbers, at least the exercise caused you to think about your instincts a bit harder. More likely, the quantitative process will raise questions about assumptions and data anomalies, which will highlight previously overlooked risks relevant to the forecast.

4. Be More Than Quantitative – Find Causal Factors

Simple statistical methods are fine for simple situations with short forecast horizons. But more variables effecting the likely forecast over a longer time horizon require the identification and isolation of factors that might cause the forecast to be greater or less than the simple statistical extrapolation. Sometimes causal factors can be obvious. For example, when forecasting anticipated growth in sales of sunglasses, one should take into account weather forecasts, as abnormally sunny or rainy weather can dramatically influence consumer purchase behavior. Other times, if you look more closely, causal relationships aren't so obvious.

The first step in identifying causal factors (if you haven't already done so), is to convene an "expert panel" of people from within your organization who possess several years of experience. Supplement the panel with suppliers, channel partners, or leading academics in the field, and ask them to identify and rank the things which tend to make sales go up or down. Try to translate the responses into definitions of factors for which there are historical measures (like weather, industry sales of complementary products, medical conditions, etc.). Where necessary, look for proxy measures that might be reasonably good approximations of the real factor – like population growth is a proxy for demand for haircuts.

Once you're identified some potential causal factors or proxies, again look to statistical methods like regression models to test the extent to which the causal factor is truly causal (e.g. is directly or inversely related to actual historical sales). Allow the quantitative process to remove any personal bias about which factors might be most causal. Also allow it to eliminate causal elements which are linked and thereby redundant.

Many forecasting experts agree that evaluating the results from multiple forecasting approaches is indeed the best way to ensure that you have the fullest perspective on the possible outcomes. Armed with that perspective, you can apply your experience and instinct to determining the most likely forecast scenario.

5. K.I.S.S.

As with most things in life, simplicity is a virtue in forecasting. Einstein said that "things should be made as simple as possible, but no simpler." In forecasting, we interpret that to mean that an accurate and reliable forecasting process should be comprehensive enough to identify the truly causal factors, but simple enough to explain to those who will need to make decisions upon it. There is no power in a forecast if those who need to trust it cannot understand or explain the logic and process behind it.

Recognizing forecasting to be a complex human decision process is the first step towards dramatically improving your batting average and improving the accuracy and reliability of the forecasts coming out of your department.

Forecasting with Data

Having data at hand which tracks the historical performance of the subject of your forecast is the envy of marketers everywhere. There's nothing quite like the confidence that comes from projecting 10 past years of sales data into the future, knowing that your only adjustments are for anticipated market and competitive conditions. With most of your causal leading-indicators (i.e. housing starts, interest rates, consumer confidence, wheat futures prices, etc.) already defined, you are free to focus on the development of contingency scenarios since you don't have to worry about the basics of who will buy this product and how much they will pay for it.

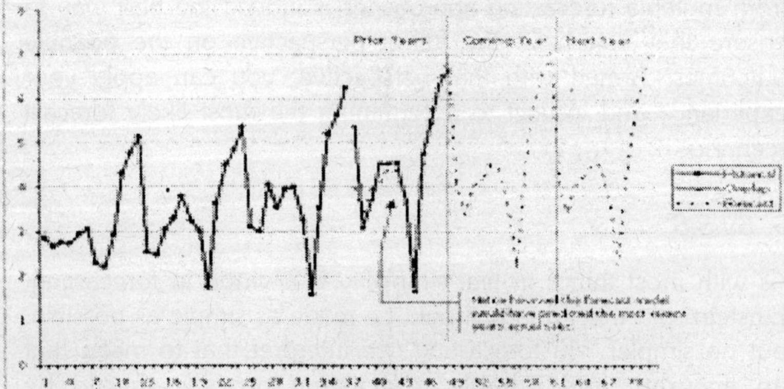

The type of forecast you are making and the number and nature of the causal factors will help determine which of the many statistical techniques are most appropriate to your forecasting challenge. You don't need to know the merits of regression, exponential smoothing, Box-Jenkins, or other statistical methods. What you do need is a Ph.D consultant or university professor to test a broad range of methods against your historical data to determine which method(s) are most accurate and/or most practical for your forecasting needs. Here too you have an advantage in that you can use the data from years 1 through 9 to feed each model and see which would have most accurately predicted year 10.

Once you have selected the appropriate statistical methodology for your forecast, you can choose from numerous inexpensive PC-based forecasting tools that can crunch the data fast, speeding up your forecasting process. The benefit of carefully selecting and then sticking with an automated software tool is that you begin to build consistent forecasting processes and measurement benchmarks. You don't have to be a rocket scientist to select and use a tool.

Today's forecasting tools are built to be used by decision-makers, not quants. For the most part, they have friendly interfaces and drag-and-drop actions to run the program. There are some are some 40-plus desktop forecasting tools on the market that range from simple excel plug-in modules to sophisticated software packages, priced from $50 to $5,000-plus. But don't expect a "plug-and-play" experience. These tools all require some degree of learning curve and familiarity with statistics. If you're just starting out, you might want to stick with the basic spreadsheet approach.

Spreadsheets vs. Forecasting Packages

Spreadsheet applications (like Excel) are found on more than 40MM PC's worldwide and are very familiar to most business executives. Excel plug-ins are also cheap (mostly under $200) and tend to be easy to use. However, spreadsheet "documentality" can be difficult because all the formulas are hidden. The cell Comments and Auditing Tools can mitigate this problem, but errors often slip through anyway. Also, scalability can be a problem as the complexity of the forecast demands grow over time. This is awkward at best, and a show stopper at worst.

In general the challenge with spreadsheets for forecasting is that they are tough to maintain and adapt. And because they are mostly run by a single-user, spreadsheets inhibit the knowledge-sharing process. But you have to give them serious consideration as a tool that most people in the office already have and are familiar with, thereby avoiding both the cost and some of the learning curve.

Forecasting Without Data

When it comes to preparing a forecast for a new product where there is no historical or corollary data, pure analytical techniques don't help much. So what do you do when you have a new product/service idea and are trying to figure out how much of it you can sell? The answer is to develop a disciplined, methodical process that best identifies and approximates the many influencing factors and market conditions facing your product launch. Here are a few approaches to consider:

Delphi Method

This method utilizes several expert panels (functional or cross-functional; preferably a mix of insiders and outsiders) who are asked to provide a forecast. Each forecast is weighted equally and the results are shared between panels, who then continue the discussion in an effort to fine-tune their forecast with points raised by other panels. If the panels' forecasts begin to converge, a final forecast can be drawn from the average of all the forecasts.

BASES Model

If you have a new product to be distributed in supermarkets,

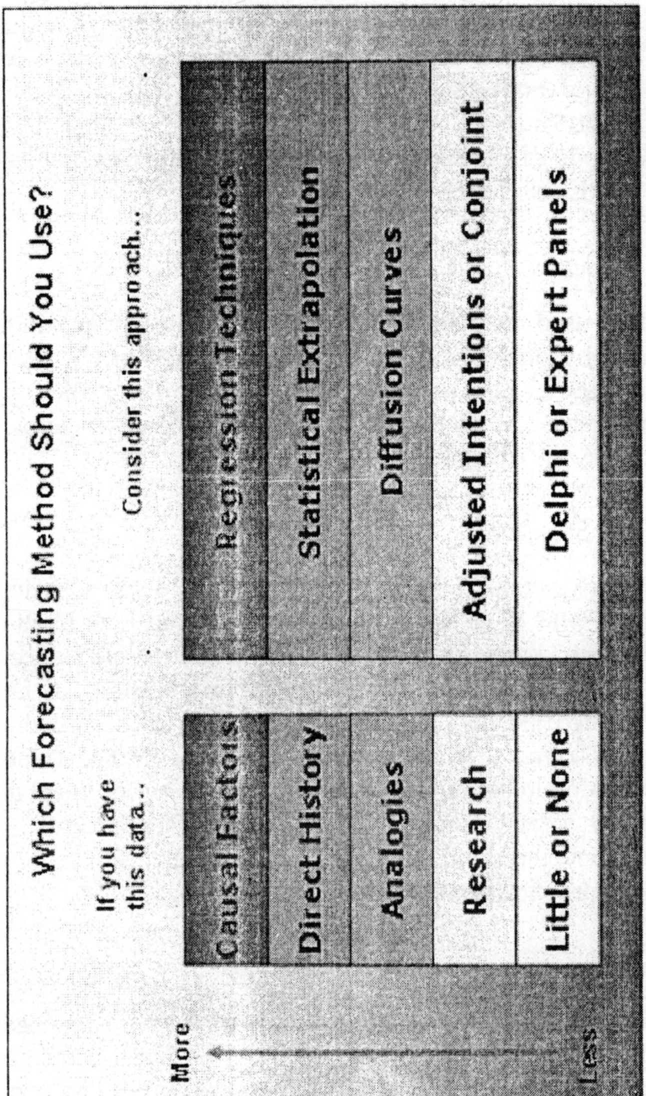
Which Forecasting Method Should You Use?
If you have this data...
Consider this approach...
Causal Factors
Regression Techniques
Direct History
Statistical Extrapolation
Analogies
Diffusion Curves
Research
Adjusted Intentions or Conjoint
Little or None
Delphi or Expert Panels
More
Less

BASES Model

drug stores, or discount stores, you might find that ACNeilsen's database of 25 years of product launch history might prove to be a good starting point. Called BASES, the methodology looks at 1000's of products with similar characteristics, consumer receptivity, and marketing plans and forecasts sales at under flexible sets of assumptions. Unfortunately, BASES is limited to just these kinds of products. It doesn't help with new paints, sporting goods, electronics, etc.

Diffusion Model

This widely used mathematical model uses rates of product adoption and usage spread from other established products to predict the rate of new product acceptance into the market. It is very applicable to both durable and non-durable goods across a wide variety of categories.

Monte Carlo Simulation

This analytical method, used in conjunction with a spreadsheet, is a form of "what-if" analysis. The name is derived from Monte Carlo, Monaco, known for its casinos and games of chance. Monte Carlo simulation is based on the probabilities of an outcome, much like the probabilities of rolling dice, and the predicted outcomes that are expected over time. Monte Carlo methods randomly select values from a defined set of possible outcomes to create scenarios of a problem. The process is repeated many thousands of times (via computer simulation) to establish probabilities of various outcomes.

Agent Simulation

Monte Carlo on steroids, Agent Simulations are highly advanced statistical models which identify and isolate the likely behavior of "agents" who are integral to the success of the product (e.g. consumer segments, competitors, distributors, etc.) based upon sets of rules found to explain past behaviors. Used for everything from air traffic modeling to war games, these sophisticated tools are very insightful where there are a large number of variable behaviors among a moderate number of agent parties.

These approaches can be combined and supplemented with market research to develop a custom process which best suits your needs. Be careful to check your preferred approach with someone experienced in new product forecasting before you go too far down the road to ensure your efforts are sound before you invest too much of your time, money, or reputation.

SUGGESSIONS FOR PREVENTING COMPETITION

Almost every hotel today has a network in place for staff use. At a minimum, your network provides "terminal" connectivity to operational applications (such as PMS), and most properties also have e-mail, Internet access and ideally, some ability to store and share files centrally. More than likely that network has been installed for a long time, with little maintenance and support. In today's increasingly risky computing environment (viruses, hackers, and reliance on computers for all functions) a lack of proper maintenance can be a recipe for disaster.

It is extremely important that the guest network and the staff network are two separate networks with as little opportunity for interaction as possible. Aside from the possibility of guest equipment or usage affecting hotel systems, security becomes more challenging if the network is shared. This is because network traffic going to and from a PC is actually going to and from every PC on the same physical network. You may wonder then, why your computer only "sees" data meant for it and not every site being accessed by everyone else on the network. It's actually the network card in your computer that is ignoring the traffic on other computers connected to the network.

There are two devices that will keep most others from seeing your data traffic: a router and a switch. A router can be used to separate two networks (guest and staff) by "routing" traffic between the Internet and other data sources to one network exclusively. A switch is actually a network hub that acts as a mini-router and routes data only to the one port (computer) that requires it. Although both these devices, when properly configured, can make the network more secure, if both staff and guest computers are physically connected anywhere, there is a greater risk. In the end, the safest course is to have a reliable networking company install and maintain your network and have the guest Internet connection separate from the hotel's administrative Internet connection (which can usually be a DSL or cable connection and even both).

Networks, computers and data are like cars. Periodic maintenance, replacing worn parts and insurance can go a long way towards averting (and allowing quick recovery from) disaster. Too often servers and software are setup and installed, then left alone until something malfunctions. The practices of data backup and updating virus software may begin enthusiastically, but eventually become secondary, "time-permitting" activities. These policies (or the absence of policies) quickly lead to risk and inefficiency and in the worst cases, lost data, an inability to service guests, and potential for business failure.

Like a car, what you can fix and maintain yourself depends a great deal on the knowledge and available time of your existing staff. Even the best self-trained mechanics are not likely to replace a head gasket however, and replacing your key server is the hotel equivalent. Clear policies need to be developed for the following activities:

— Installing and maintaining virus protection software on all computers

— Updating all hardware drivers for computers and network equipment

— Backing up key data daily and ensuring that at least some backed up data is stored offsite

— Installing all critical operating system software updates, including database software, as soon as it is released by the vendor.

The Rules

Some of thóse who thought HR rules were silly and obvious have learned the hard way that the cost of establishing, maintaining and enforcing personnel policies is small compared with a single employee abusing the unstated policy. For the same reasons, policies for data security, appropriate access and other aspects of system use are critical to preventing and responding to problems that arise. One defense that hackers have successfully used in the past is that they were not aware that they were not allowed to access a computer system, simply because there was no notice advising "Authorized Access Only".

You certainly want to consult with your HR, legal and perhaps risk management advisors, but establishing policies on things such as passwords, permissible data access, protection of guest data, and the hotel's right to monitor employees is a "must do" item. This will make it easier for employees to comply and for the hotel to respond appropriately to any who don't. By working with your software suppliers and interface vendors, you can ensure the security is established at an appropriate level and that third parties with access to your data (as well as some responsibility for protecting it) are also maintaining an equal emphasis on security.

Virtual Private Networking

A VPN is a simple ways of increasing security while using a public network whether it's wireless, a standard wired network, or the Internet. The TCP/IP protocol and Ethernet standards pass data from many computers along the same network hardware. What keeps you from seeing other user's data does not stop others from seeing your data, provided they have the right equipment and desire. What a VPN adds to the equation is simple encryption. The data being sent and received over the network is encoded so it's not easily readable by anyone who does intercept it. Since your workstation or laptop is actually encrýpting and decrypting the data it stays "unreadable" until it reaches its destination.

Unfortunately, the VPN requires that your computer is talking to another computer enabled by a compatible VPN. This requires a more sophisticated server or router, but it is well within the capability of most networking companies. Another consideration with VPNs is the necessary processing power for encrypting and decrypting all data. For a workstation or laptop computer purchased over the last few years this will likely not be a problem; however, for a server that may have to manage VPN connections with many workstations or other computers, this may place too great a load on the system. Establishing a VPN for staff use in the hotel has other considerations that may make it impractical for a smaller property, but the added security is worth investigating.

Networks should be thought of as communications media. Although their communications revolve primarily around data, they are little different than phones, televisions, radios and printed matter. Similar to other media, some communications are designed to go to anyone and everyone, while other communications are private. An appropriate analogy might be that e-mail is like a phone call and your web site is a combination of TV, radio and print.

The equivalent of a knowledge library, conference call and closed circuit TV rolled into one can be likened to an intranet. An intranet is simply a browser-based (web site) application, restricted to authorized people. The access that each person or group has may be different, or it may be that anyone who is allowed on the intranet can access all of its content. An intranet is a wonderful communications tool that is only now beginning to make its effect felt in many progressive organizations.

The intranet can provide basic employee information such as a directory, frequently requested forms and policies. It may have employee-specific data, such as schedules and secure intra-employee e-mail. It can even offer a full-blown EIS (executive information system) where managers have access to data, reports and analysis without having to page through monthly reports that are sometimes outdated as soon as they are circulated. Similar to an intranet, but including access (again, only for those authorized) for other organizations (such as suppliers or channel partners), is an extranet.

Like an intranet, it is browser-based and accessible by authorized users from any Internet access point. Because both intranets and extranets use passwords for anything but very general (and hence, public) information, VPN access is advised. Technology crosses virtually every area of property operations, sales & marketing and management, and its necessity, power, and complexity are only increasing. And for those who have relied on "tried and true", it's time to recognize that there is no such thing when dealing with a phenomenon that grows at over 100% each year.

A crash course in hotel acquisitions

Hotels are a business. Real estate considerations such as location, market, and physical condition are all important but if you miss the business issues, the deal will go bad. Over the last eighteen months HVS has been approached by and worked with numerous investors who are looking to place money in the lodging industry. Most of these investors are experienced and savvy commercial real estate investors but are novices to lodging investments. The general wisdom is that both the current pricing and economics are right for making deals and that returns will be significantly higher than other forms of commercial real estate. The novice hotel investor will soon learn how hotels are different from other forms of commercial real estate. Here are a few tips to help them on their way:

Hotels are a business: Real estate considerations such as location, market, and physical condition are all important but if you miss the business issues, the deal will go bad. Suffice it to say that the management company you select is one of the most important decisions you make. Strong management can make average deals great, poor management can make great deals bad. In the hotel business your customer is the general public and the right franchise company can help you reach your market. Hotel rooms are rented on a nightly basis to hundreds of individuals, as opposed to the office and retail sectors where a handful of large single users sign leases for multiple years. These experiences and relationships must be expertly analyzed and managed or revenues and values will suffer. Assuming that you have made the correct decision regarding the above two relationships, what are the other business issues that will have to be considered?

Let's start with employees: A small, limited service hotel can have 20–25 employees while a full service hotel can have hundreds. Again, this is dramatically different from most other commercial real estate assets that can be operated with a handful of maintenance personnel and a small office staff. Most often the hotel staff are employees of the management company rather than of the owner. However, the owner needs to be certain that human resource issues are handled correctly to avoid potentially costly liabilities.

A review of the current sales and marketing plan for the property can disclose issues that will have an impact on future revenues of the hotel. Frequently, hotels sign room rate or advertising agreements that can have a bearing on the performance of the hotel for a year or more into the future. New owners' need to fully understand what sales directions have been set for the business and how long it may take to change course, if necessary. The adequacy of these plans also should be fully understood. Often times the purchaser will assume that their team can do a better job selling the property.

Hotels also differ from other commercial real estate in the level of personal property and operating inventories that they maintain. This is true for limited service hotels and even more dramatic for full service hotels that maintain food and beverage inventories. These items can lead to governmental permitting and licensing issues and can also result in sales tax liabilities. Some municipalities require the purchaser to pay sales tax on the value of all personal property at the property at the time of purchase.

The purchase and sales agreement must detail how these issues are to be handled. What are acceptable levels of inventories and supplies at closing and how will discrepancies be handled? Without a thorough inventory the purchaser can be left with a hidden funding need at closing that can easily amount to tens of thousands of dollars.

Hotels also differ from standard forms of commercial real estate in the number and types of maintenance/service agreements and information technology requirements. All of the documents relating to these items must be reviewed and in many cases new contracts must be negotiated. While many of these agreements will be routine for the commercial real estate investors such as alarm monitoring, boiler maintenance, elevators maintenance, etc., many are unique to the lodging industry. These include television agreements, telephone call accounting systems, credit card processing, check processor, pay per view movies, restaurant point of sales equipment, property management systems, back office accounting systems and procedures, vending agreements, and linen service.

The issue of closing adjustments that must be addressed also differs for hotels. These adjustments should be fully detailed in the purchase and sales agreement. In addition to standard adjustments, hotel specific issues will include:

— How is cash on the property handled?
— Who gets credit for the revenues from the day of closing?
— When are employees turned over to the new employer?
— How will the City and Guest Ledgers be handled?
— How will F&B Inventories be accounted for?
— How will discrepancies in consumables inventories be handled?

The novice lodging investor needs to carefully understand the unique aspects of hotels and make informed decisions during the purchasing process. With adequate knowledge and planning, returns can be achieved that will exceed those available in other forms of commercial real estate.

Updating Web site

When it comes to updating your web site, the first and most obvious question your web team must ask themselves is – "did the site meet, surpass or fall short of the objectives that were set out last year?" In order to answer that question, you must have previously established clear, quantifiable objectives. You would be surprised to learn that there are still quite a few relatively sophisticated site development teams out there that fail to quantify their objectives into terms that can be measured and benchmarked.

And as the saying goes, you can't manage what you can't measure and when it comes to the web, it is imperative that you are able to quantify your site's performance against objectives and benchmarks.

Effective User-friendly Website

A web site in most cases can be a useful, cost-effective tool for achieving your objectives, but it is not an objective in and of itself. If you are not clear why you are investing in the development of this unique channel, it will be near impossible for you and your team to create a site that is user-friendly and builds long-term relationships. In the hospitality sector, building long-term relationships is critical if your site is to produce an acceptable return on your investment. The pool of potential visitors to your site might seem bottomless, but in fact, there has been significant research to indicate that once users abandon a visit because of a negative interaction, there is very little likelihood they will soon return.

In other words, above all else, your site must incorporate a "user-friendly" personality that begins the moment a visitor types in your URL. Even something as simple as the URL you choose when registering your site can contribute to a sense of user-friendliness. You can try and be creative, but most people when searching the net will enter www.(your hotel name).com or use one of the more popular search engines like Google or Overture to find you. Do not throw in unnecessary words or dashes to complicate the URL. Keep it short and true to your name. Your users will appreciate it.

The moment a visitor arrives at your site is when the proverbial "rubber meets the road". All your hard work and energy, your investment in top quality art and photography, the millions of dollars in technology and design and the people hours invested to insure your visitor has a pleasing experience will mean nothing if your home page takes more than 20 seconds to open.

You may have great high-resolution photos of your property or a trip-planning calculator, but if it takes too long for them to open on your visitors PC, you have wasted your money and your visitor's valuable time. Even more important, you have an unhappy customer on your hands. Remember when you are designing your site that not everyone is accessing it from the office with a high-speed broadband connection. One option you might want to consider is to let your users use a version of your site based on how they are connected to the web.

For example virtual room tours, videos, high-resolution photos, music and ambient sounds etc might be offered on broadband versions only, while dial up users are offered features tailored to the download times of their connection. In order to make a site user-friendly you have to cover all the basics.

Content is the next dimension you must be on top of if you want to insure you have a User-friendly site. You must be able to measure your user's level of satisfaction related to the overall content found at your site. Are you giving them enough information to make a reservation? Are you giving them enough detailed pricing information? Is the information on your site current and does it accurately reflect your brand. Your web team must have this information at their fingertips on an ongoing basis if your site is to remain user-friendly.

The one constant on the web is that it is constantly changing. A feature your users liked and took advantage of last quarter may be old and outdated next quarter. Listening to your users is an ongoing exercise, not something that is done once a year. Another basic element in every user-friendly web site is Interactivity. You must insure your users are able to interact with the site. Tools like help buttons and site maps are mandatory if your site is to be perceived as user-friendly.

Giving your users contact information is another way of demonstrating that there are real people concerned with satisfying their needs and ready to interact with them if needed. It goes without saying that if your organization is going to offer e-mail as a communications channel, it must be prepared to respond to e-mail within a reasonable time frame. From a net perspective, anything longer than 24 hours is not acceptable. Enhanced Interactivity is a key factor in building user trust, which in the experience is by far the most important attribute related to enhancing user loyalty.

Being user-friendly also means providing your users with a reason to return. Motivating them to use your site as a resource and making it a first stop when they are seeking hospitality related information. Providing motivation can be as simple as helping them save a little time with access to a calendar for example. Constantly adding new features, services and content to your site is also a great way to insure people are motivated to check back in to see what's new. Having said that, you need to monitor their feedback, preferably in real-time to make sure your new features for example are not making it more difficult for your users to navigate throughout the site. Waiting months for a research report to indicate there is a problem is not a good idea.

User-friendly is a state of mind. It is not something you achieve by offering your visitors less of something or more of something else. It is about listening to and understanding the needs and wants of your users and aligning those needs with your objectives.

HIGH-SPEED INTERNET ACCESS

With the growing demand for high-speed Internet access in

hotels by the traveling public – most notably business travelers – hoteliers are looking more closely at how offering such a service can be beneficial to their property and their guests … and the revenue implications it may have. As high-speed Internet access has become a more popular offering for hotel guests, hotels are investigating – and rightly so – all their options for providing such a service.

An increasing number of hoteliers have begun considering and some even implementing high-speed Net as a free amenity to their guests. Though a free-to-guest model may be appropriate in some cases, it may not be the best option available for the hotelier or their guests. There are a number of issues that should be considered when deciding to charge or not to charge for high-speed Internet access. With a comprehensive view of the entire issue, hoteliers will be able to make the best decision for their unique situation.

A recent survey investigated business travelers' attitudes toward high-speed Internet access in hotels, including awareness issues, usage factors, price sensitivity and other relevant factors. Internet access users in hotels fall into five general segments:

1) high-speed believers (use high-speed Net whenever available);
2) rational choosers (use high-speed Net when it fits their needs);
3) newbies (not highly familiar with the two different options);
4) price-constrained dial-up users (use dial-up when high-speed is deemed too expensive; and
5) die-hard dial-up users (use dial-up all the time, regardless of availability or price of high-speed).

The high-speed Internet access has become a must-have for many travelers, and that many more are poised to become regular users of the service as well. It also uncovers the interesting – and perhaps counterintuitive – point that price is not a major factor when it comes to usage of high-speed Internet access. Most travelers expect to pay for the service, and price point is not the major factor for them in deciding whether to use it or not; this likely is largely due to users being able to expense the charge as a business-related cost.

Free-to-guest may be a good option for a limited-service hotel that doesn't anticipate extensive usage of the service or feels that a potentially lower quality of service is acceptable for guests. And while there may be sound reasons for offering high-speed Net for free, it should be understood that there also are challenges of doing so ... challenges for the hotel, for the service provider and potentially for the guest.

First, under a free-to-guest model, there will be pressure to provide the service at the lowest cost possible, in which case the service itself may suffer. For example, if high-speed Net is positioned as a free amenity, it may consist solely of hard-wired connectivity as provided by the hotel on its own – without the type of comprehensive solution including professional network engineering, system monitoring, sales and marketing, customer care and much more that a full-service provider can offer.

Secondly, the costs of providing high-speed connectivity (i.e., bandwidth charges, etc.) are likely to rise when the service is free, assuming that the attractiveness of "free" will drive at least some usage. This is the type of problem that will only compound itself as usage rises, since you'll be adding no revenue but costs will continue going up.

The bottom line is that the decision to charge or not to charge comes down to your own unique situation. Deciding what is best for you and your guests in terms of high-speed Internet access involves understanding your guests' needs and expectations, while matching those up with the opportunities you have to benefit from them. Ultimately, the goal of amenities such as high-speed Internet access is to serve your guests and potentially drive occupancy; how you choose to do so remains up to you.

Investing in Hotel Portfolio

During the last three years, when it comes to investing in hotel portfolios, the pace has picked up and more players have entered the mix according to Jones Lang LaSalle Hotels' research report titled *FocusOn Hotel Portfolio Sales*. Moreover, the binge-buying of the mid and late 1990s has left a number of hotel investors with the burden of holding non-strategic assets, or misfits. In the current environment of very low interest rates, low nominal returns, and the hunt for yield, it is essential that portfolio investors become proactive in diversification and balancing.

Common among hotel owners is the desired holding period and the willingness to invest additional capital to capture upside. Ownerships' appetite for risk, capital availability and required rate of return will dictate the long-term strategy for a property, while the exit strategy is also critical to maximizing investment returns." The entire hotel portfolio asset by asset is a major key to long-term performance. "It may be that some hotels in a purchase are valued at zero (loss-making operations or over-leveraged properties), thus offsetting some of the inherent value in the higher-quality properties. Flipping these properties out of the portfolio will immediately raise the overall yield," included Elgonemy. "Other essential considerations for sellers include whether the portfolio is financeable, whether management is available or not, and the extent of deferred maintenance."

When it comes to debt considerations at the portfolio level, hotel lenders seek to put money into the market, even as they continue to be very selective. "The amount and pricing of debt financing is critical to maximizing the return on equity and increasing the diversification possibilities within the real estate portfolio. Also, maximum proceeds can be obtained by cross collateralizing the portfolio.

Portfolio sales enable buyers to achieve strong savings and efficiencies that are passed through to the properties and create opportunities for large investment funds to immediately expand in size and geographic location, leading to intensive competition during the selling period and higher yields. The expertise of undertaking a portfolio sale compared to selling several single assets at once saves time, money and administrative energy. It is an advantage that will encourage more sellers to enter this evolving arena in the future.

Jones Lang LaSalle Hotels, the world's leading hotel investment services group, provides clients with value-added investment opportunities and advice. In 2002, its success story includes the sale of 6,747 hotel rooms to the value of US$862 million in 36 cities and advisory expertise on 116,877 rooms to the value of US$17.8 billion across 170 cities. Jones Lang LaSalle Hotels' services include transactions, mergers and acquisitions,. financial advice and capital raising, valuation and appraisal, asset management, strategic planning, operator assessment and selection and industry research. Jones Lang LaSalle (NYSE: JLL) is the world's leading real estate services and investment management firm, operating across more than 100 key markets on five continents

Optimizing user feedback in the hospitality sector

In the hospitality sector specifically, a web site can be one of the most valuable weapons in your marketing arsenal for listening to, understanding, and responding to the needs and wants of your customers, both online and offline. Your site could and should encourage feedback from your users. It makes sense to hear what people have to say. The more you listen, the sooner you can create a user experience that really connects. Feedback is particularly important for the marketing team working on your site. They need to listen carefully to their audience. They need to understand what your users want and what they need in order to understand what upsets or pleases them.

At the end of the day the key objective of most web sites in the hospitality sector is help boost revenues as an alternate reservation channel or to support and enhance the organizations marketing and advertising efforts as well as strengthen the organization's relationship with its customers. And because relationships are dynamic, what web site users need today to build a strong relationship with your hotel may be different from that of tomorrow.

By continuously listening to and responding to what users like and don't like about your property and/or its e-commerce media, a strong relationship is created that turns users into repeat customers and repeat customers into loyal advocates - in effect, your users become "in the trenches" visionaries for your hotel as well as advance scouts and pro bono consultants so to speak. The key to building such loyalty is not only discovering these previously invisible connectors, but also knowing how to use them to increase the bond between the organization and the user. The deeper the bond, the greater the value each brings to the relationship, and the stronger the brand.

When organizations regard users as long-term assets, they maximize this exchange of value. By listening to users, responding to them and collaborating on what the relationship means from their point of view, an organization not only increases the web site's ROI, it gains other valuable benefits such as user advocacy, brand loyalty and trust. With so much at stake, it seems strange to me that so many hospitality organizations pay such little attention to collecting feedback from their visitors.

If you are a smart businessperson you realize that by listening to what your web users are saying you can quickly learn who your web visitors are and collect valuable demographic information about them such as the purpose of their visit and how often they travel etc. For each of the 'communities' of your targeted customers, you can learn about their likes and dislikes and their level of overall satisfaction with your site and even your hotel.

User feedback is a valuable business intelligence tool that no web site owner in the hospitality sector can ever afford to ignore. A site organized by functional departments is usually useless to customers who don't care what department something's in. They just want a problem solved. On the other hand, a highly creative site might be completely wrong if your customers are looking for comparative room data, or interactive assistance in making a reservation. The web offers you a unique opportunity for taking the pulse of your present guests and prospective customers. Something that is prohibitively expensive in the real world.

On the web, technology allows to automate the process of collecting user feedback so that each and every customer gets a chance to be heard — at an extremely low cost compared to traditional marketing research methods. With the growing influence of "users" in today's on-demand society, hotels must learn how to adapt and continually satisfy the need of their users if they wish to remain competitive and maintain or grow their share of market. Customer feedback can be used to build and maintain a relationship between a hotel and its stakeholders and in fact, understanding, interpreting and correctly responding to this feedback is the key to successful business and solid branding. It is about a promise of assurance and a perception of future excellence that leads to establishing a solid customer relationship and ongoing user loyalty.

In the early days of Internet marketing, the goal of most web site owners was to drive as many potential customers to their site as possible and hope that some of these visitors would conclude a transaction, thus creating a new and relatively inexpensive channel for generating revenue and reducing costs. And this will not dramatically change until site owners recognize that to get users to make an online reservation, the site must first and foremost establish a bond of trust with them.

On the web, establishing a bond of trust is the first and most essential step in creating an ongoing, successful and profitable relationship with your user. Online business is conducted at arms length. There is no front desk staff to speak with. No reservation person to question. No concierge to provide assistance. There are no physical buildings to touch. No pool to swim in. No dining room to eat in. There are just pictures, images and words. In a nutshell, if you have not established a previous level of sufficient trust with your web site users or your web site itself does not create an environment of trust, visitors will be extremely cautious and hesitant to share information or commit to making a reservation.

Trust is the barrier that users must cross on their road to making a reservation. Trust is rated significantly higher for users coming back to a site to make a reservation than for those simply at the site looking for more information or comparing rates. It is interesting to look at the various ways users interpret "trust" across hotel web sites. Trust encompasses believing that the room rate quoted on the site is the lowest or best available for that time period. It means trusting that room descriptions and pictures are accurate and fair representations of the actual physical make-up of the property. It means trusting that availabilities are accurate and that reservation confirmations will be honored. Finally it means a site will also maintain a high level of privacy and confidentiality with the information that the user has shared with the web site owner.

First and foremost is the fact that users are more likely to share personal information - the data you need to successfully build a relationship with them. Using this data, web site owners can provide a more personalized, productive and time saving experience for the web site user increasing their level of user satisfaction which in turn should result in higher "look to book' ratios", lower acquisition costs, increased revenue and of course increased profits for the site owner. This also translates into a significant competitive advantage.

Direct Online Distribution

To begin with, there is a new realization among hoteliers that something has to be done to change the status quo in online distribution. Hoteliers are becoming more and more frustrated with the fact·that they are consistently losing leisure and business customers alike to the online discounters.

For many hoteliers Direct Online Distribution has become a top priority. In the offline world hoteliers enjoy more direct sales than intermediary sales. In the online world hotels are less aggressive than the airlines about bypassing the intermediary channel. This year for example only 52%-53% of online hotel bookings will be direct sales. Unlike the airlines, hoteliers have difficulty maintaining market share and finding the right formula to deal with online intermediaries. The lack of comprehensive Internet strategies, ineffective online distribution efforts and the explosion of the "merchant model" are partly to blame for the current situation.

Very few hospitality companies were able to take full advantage of the Internet as the most efficient distribution medium over the past year. Many hoteliers continued to suffer from their lack of understanding of the dynamics of Internet distribution. The online discounters took advantage of the situation and increased market share at the expense of hotels' direct and traditional distribution.

Hoteliers' online market share fell from 54% in 2001 to 52% in 2002 and it will take extra efforts to increase this share in the future. Hoteliers are vulnerable due to the changing dynamics in online distribution and the proliferation of the merchant model. Hoteliers' direct share will depend to a great extent on how aggressive their direct-to-consumer sales efforts will be in 2003 and beyond.

Hotel Branding on the Web

Anytime an Internet user lands on a website, a branding interaction occurs. This branding interaction can be positive (brand-building) or negative (brand-eroding). The question is, who is the Internet user interacting with and who benefits from the branding interaction?

Over the past two years hotel websites have become the main "point of contact" with potential customers. A well functioning, fully optimized hotel website is a real branding asset that serves as the chief instrument to capture new markets, entice potential customers to stay at the hotel and facilitate transactions. Surveys show that overwhelming majority of customers will not stay at a hotel with a poorly done website that does not reflect well the hotel product and the destination. It is becoming a common belief now that if your hotel website is not up to par there is something wrong with the hotel itself.

Search Engines

Search engines are considered major brand builders on the Web. Hotel listings or paid advertising on Search Engines, Directories and Portals (e.g. Google, Yahoo, MSN) contribute directly to the brand building of the hotel itself. Due to the nature of these services and their perceived impartiality, the branding interaction is exclusively between the Internet visitor and the hotel. A natural (algorithmic) listing of the hotel website, paid hotel banner ad or pay-per-click sponsored listing directly contributes to the brand building of the hotel, not of the search engines.

Online Advertising

Online advertising plays a dual role: as branding and direct response tool. The Web offers hoteliers not only great branding opportunities, but also unlimited opportunities to create interactive relationships with their customers. Email marketing and pay-per-click marketing are only two of the very potent direct-response tools at the disposal of smart hoteliers. Unlike the above scenario, pay-per-click marketing on the search engines, and consumer email marketing are ideal direct-response vehicles and powerful tools in the hotelier's direct-to-consumer online distribution strategy. Pay-per-click marketing is also a great branding tool.

The merchant model has greatly hurt the hospitality industry and has done long term damages to the hotels' brand and price integrity. You don't need a bigger proof than that, just look at the diminishing ADRs. The so called "rate parity" is in fact acceptance of the merchant discounted rates as your rack rates. Your hotel's discounted rates on Hotels.com and Expedia are de facto your hotel's published rates. Though claiming to be "free of charge", these merchant services cost hoteliers dearly. They cause long term damage and downward pricing pressures (both online and offline) beyond repair.

What can hoteliers do to avoid brand erosion?

The answer is very simple. Hoteliers must adopt a comprehensive Total Online Distribution Channel Strategy, a comprehensive online channel strategy, which turns the direct-to-consumer distribution model into the centerpiece of the hotel's Internet strategy and optimizes the balance between the Direct and Indirect Web Distribution Channel.

Online distribution is a very complex undertaking that involves a variety of online channels, services, marketing approaches and sales techniques. Therefore we call the Internet strength of a particular hotel "Sphere of Web Distribution Influence", which we define as the percentage of Direct vs. Indirect online distribution influence, presence and exposure of the particular brand. The higher the Direct Sphere of Web Distribution Influence, the less dependence on online intermediaries.

The direct-to-consumer distribution model should become the foundation, the centerpiece of any hotel company's online distribution strategy. Why? First of all, The Internet is the ultimate "Direct Distribution Medium". Why direct distribution is so important? It provides the hotel with long-term competitive advantages and lessens dependence on intermediaries, discounters or traditional channels that are about to become obsolete. Direct-to-consumer online distribution has the following benefits:

— Puts the hotel in control of its Internet presence and exposure
— Prevents brand and price erosion
— Lessens dependence on online discounters and intermediaries
— Is the shortest path to establishing interactive relationships with customers
— Provides long-term opportunities to benefit from the lifetime customer value

— Is the least expensive way to distribute hotel inventory—direct to consumer!

Direct online distribution is not just a theoretical, "ideal scenario" approach. Many surveys show that online customers prefer dealing directly with the travel suppliers, including hotels, when purchasing travel online.

Technology has liberated the industry from time-consuming paperwork and repetitive manual tasks – especially in the critical area of property management systems. 25 years ago, many hotels used the Whitney Board system, with a rack full of colour coded slips of paper to show whether a room was occupied, in need of cleaning or ready to sell. Simple reservation systems were maintained manually except those coming through a central reservation system and telexed overnight. It took hours to complete the night audit, and any guest wanting to check out before it was completed would have been met with a curt response.

Today's hotel executives work in a completely different environment. Relative simplicity for both guest and hotelier has been replaced by either complete anarchy or total choice depending on where you are in the value chain. Here, we suggest that hoteliers need first to find the value drivers for their unique business, and then bring together the most suitable elements to create the ideal property management solution for their organisation.

There are four key technology elements that enable executives to assemble a solution that is flexible, yet meets the appropriate needs of their unique customer proposition.

First, the ubiquity of the *Internet* as a communication channel both between the hotel itself and all members of the extended enterprise (corporate sales, central reservations, headquarters) as well as between the hotel itself and all of its direct (business and leisure) and indirect (travel agents, tour operators, conference and convention planners, consolidators) customers.

Second, the emergence of *open systems* (ones that adopt the Open Travel Alliance standards and/or adopt Microsoft standards) allowing CIOs to bundle the best applications to meet each business's needs rather than acceptance of a pre-bundled series of proprietary applications.

Third, the rise of the *networked enterprise*. This enables chains of hotels to reach the goal of a single image inventory, and allows even the independent unbranded hotel to determine the distribution channels for its room inventory in the future. Owners and brand operators can decide to bundle all rooms in a city and/or in a segment and treat them as if they are one single inventory – or can continue to ask unit management to drive inventory and pricing decisions.

And fourth, the adoption of 'thin client' web browsing technology that allows the application to be held on the organisations intranet server, or to be provided by an application server provider. This means that the business is no longer organised around the technology tools – with the tools becoming the servant rather than the master.

Increasingly, organisations are moving away from both these models and are following the network solution path. This means that each of the key functional groups identify the best application for their purpose and define upfront what data needs to be shared with other functions. The network is then developed with common data definitions and dictionaries, a single central integration tool and common open standard integration, regardless of source or target applications. Key processes are linked together using the enterprise application integration tool and workflow methods.

The approach enables both a presentation/executive dashboard level to be built into the architecture and a common content management system can be built in if needed. The networked approach can be developed step by step from either a centralised or decentralised system, and provides a flexible architecture with market-leading applications serving differing communities. It drives the adoption of best practice into the business, but requires the different parts of the organisation to be much less territorial. It also means the business must first define and lock onto the key value drivers for its unique target customers, and this can be a complex and difficult exercise.

The diagram in the next page illustrates the various layers of technology. The goal for each organisation is to find the best applications in each layer and then bring them together into one network. This way, the property management system's core transactional functionality is harnessed into a network with the functionality of other applications. This will benefit customers, brands and regions as well as the hotels themselves.

Different businesses have different needs and some groups consolidate functions, while others run them separately. And the ideal solution for any one company will change as the business itself develops and diversifies.

Lowest Price Guarantees in Hospitality

The catch phrase "Lowest Price Guarantee" is an age-old marketing tool widely used in retail since ancient times. Even before the Romans knew how to conquer merchants were selling their wares in bazaars and marketplaces drawing in buyers with a lowest price guarantee.

The lowest price guarantee is applicable to every hospitality business model from corporate brand to hotel management company to the independent corner hotel. The lowest price guarantee is one of the most potent techniques hoteliers have at their disposal to boost direct-to-consumer online distribution. Used wisely such a tool can play an important psychological and promotional role in online distribution. Hoteliers benefit by offering the lowest price guarantee:

— Serves as a great tool to beat the online discounters at their own game

— Such guarantees entice more consumers to book from you online

— Rate guarantees can dramatically boost conversion rates on the hotel website

— It is less expensive to guarantee the lowest price than to sell through online discounters
— Rate guarantees are great customer retention instruments
— Such guarantees have important psychological and promotional effects for consumers
— More and more, the guarantee is an expectation by online consumers
— Price guarantees have a great "word of mouth" effect.

The following two key considerations clearly support the importance of the lowest price guarantee:

— Online travelers (lookers and bookers) claim that "assurance of lowest price" will influence them to book, more often, online.
— Online travelers overwhelmingly prefer dealing with travel suppliers over intermediaries.

In other words, by providing a lowest price guarantee, the travel supplier responds in a very categorical way to both primary preferences of the online traveler: guaranteed assurance and supplier direct.

Wireless in Hospitality Industry

As with all businesses the travel industry is fiercely competitive, and the hotel business is no exception. Choosing which hotel to stay at can be a daunting and confusing task, there is the price, location, services, add ons, promotions, quality, and other factors all to consider. This is one reason why hotels love gimmicks, and as soon as one property has made the move and introduced a new service, which may attract customers from one hotel to another, you can be sure that most other hotels of a similar status will follow suit and also add that service.

Such gimmicks, include things such as Frequent Flyer points, membership programmes, separate showers, numerous telephone lines, mobile telephones, executive floors, complimentary services, tea and coffee making facilities and the list goes on and on. Now in 2003 the latest craze is wireless connectivity. The ability for guests to connect to the internet from anywhere whether they are sat at the bar, relaxing in the lounge or waiting in the lobby, no matter where you go, the hotel wants you to be able to stay connected.

Some gimmicks do not always work as hotels expect, take in room broadband access as an example. A mad craze swept the industry as hotels rushed to install broadband connections in their rooms, only for many hotels to later discover that only roughly 10% of guests use the service. One reason is perhaps the fact that frequent travellers going from one city and / or country to another quickly learn that each hotel/city/country uses a different system requiring them to often reconfigure their computers to use each system.

Wirelesss connectivity is not of course limited to laptop users, the latest mobile communication devices can also be plugged into these wireless systems, and these users will most probably gain the largest benefit from being able to connect from anywhere, for one simple reason - Power. Hotels will often tell you about all the wonders of their new wireless system, but when you ask about where users, especially laptop users, will get their power from, many hotels draw a blank. The average laptop user will only get 1 - 2 hours from their computer battery, and so for many having a power source would add a great deal of weight to the ability of using the laptop anywhere in the hotel.

Not every hotel has overlooked the power issue, the InterContinental Hong Kong for example, in its amazing new Executive floor lounge has, if you manage to make it past the complimentary Dom Perignon, installed well concealed power points around most of the seats. The number of hotels, even properties which are still under construction or just about to open that seem to have overlooked the question of power seems to be in the majority, and without power, we see little point in going wireless.

INTERNATIONAL HOTEL INDUSTRY

While there are difficulties about the precise definition of a hotel, at the end of 1998 the latest statistics available showed that there were 15.4 million hotel rooms worldwide. Since 1990, the industry has recorded an annual growth rate of 2.9%, which gives approximately 16.3 million rooms by the end of 2000. Worldwide supply of rooms is far from evenly distributed, however, as shown by figure 1.

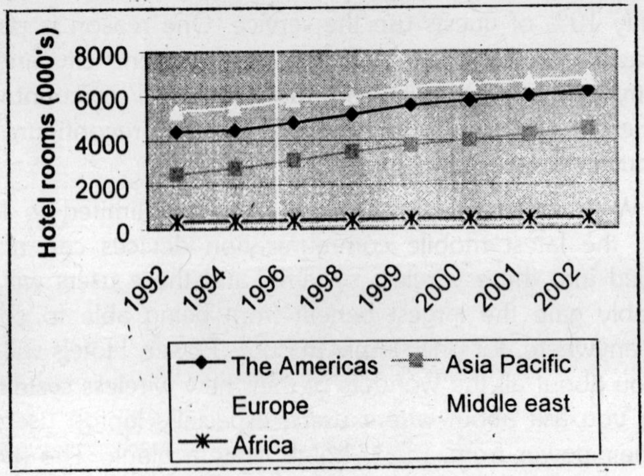

Figure 1. Number of rooms in hotels & similar establishments worldwide, 1992-2002

Although the majority of hotel rooms, as shown above, are still within Europe, the rate of increase has been much lower in Europe than in the other two major regions of hotel supply, the Americas -principally North America - and Asia: North America now has over 4 million hotel rooms, and Europe over 6 million. Although the latter remains the dominant destination worldwide, over the next twenty years it is in Asia, (and mainly South East Asia) that most tourist growth will occur. The overall share of hotel rooms by continent can be seen in table 1.

The trends are clear. Europe and North America, which started the decade with a combined market share of over 70%,

Table 1. Worldwide hotel room capacity % share by region, 1990-1998.

	1990	1992	1993	1994	1995	1996	1997	1998
The Americas	35.2	35.5	36	35.4	32.9	33.1	33.3	33.5
North America	29.9	31.7	31.1	31.2	27.5	27.4	27	26.8
Caribbean	1.1	1.3	1.3	1.3	1.2	1.2	1.2	1.3
Central America	0.3	0.3	0.4	0.4	0.4	0.4	0.4	0.4
South America	4	4.2	4.1	4.3	3.9	4.2	4.6	5
Asia Pacific	20.5	20	20.6	20.3	22.5	22.9	23.3	23.7
Oceania	1.6	1.8	1.6	1.8	1.6	1.7	1.5	1.5
Europe	40.2	40.3	39.3	40.1	40.4	39.7	39.1	38.5
Middle East	1.3	1.4	1.3	1.3	1.3	1.4	1.4	1.4
Africa	2.7	2.8	2.8	2.9	2.9	2.9	2.9	2.8
Worldwide	100	100	100	100	100	100	100	100.

Source: TTI, 2001.

Table 2. Sources of business for hotels 1992-1999 % of total room nights

Source Of Business	1992	1993	1994	1995	1996	1997	1998
Domestic	47	48.1	49.2	51.4	53.5	47.4	48.9
Foreign	53.9	51.9	50.8	48.6	46.5	52.6	51.1
Composition of Market							
Business Travellers	33.6	36.1	38.3	36.9	34.3	28.5	31.9
Leisure							
Travellers	42	40.5	22.3	22.7	26.8	34.1	33.8
Tour Groups	N.a.	N.a.	17.3	18.4	16.5	17.7	16.1
Conference participation	11.3	10.3	9.9	10.9	9.2	10.1	9.8
Government officials	3.9	2.9	3.9	3.1	3.3	2.2	1.8
Other	9.2	10.2	8.3	8	9.9	7.4	6.6
Total	100	100	100	100	100	100	100

Source: TTI, 2001, p.13.

had less than two thirds of the market just eight years later. Some regions of the world have remained largely static in their hotel supply, for example the Middle East and Africa, while there has been a low level of growth in South America. It is striking that Africa still has less than 3% of the supply of hotel rooms. This points to the lack of development in the tourism industry, and perhaps to the potential for growth in future. There are signs that countries such as South Africa and Namibia are starting to develop their tourist industries more fully. Such development remains highly dependent on stable political and economic conditions, however: the record of the industry in Kenya, for example, shows how vulnerable it can be to negative publicity.

Composition of the market

Development of the hotel sector is due to a healthy tourism industry attracting international leisure visitors, coupled with an expanding economy that stimulates demand for business and leisure through domestic demand and in bound business tourism. Table 2 gives a breakdown globally of the sources of business for hotels, and the major market segments of demand.

With regard to the first part of the table, which concerns domestic and foreign business, it is difficult to explain the reasons for the changes in the rate of increase in domestic and foreign business. Until 1996 there is a consistent gravitation towards domestic business, which by the end of the decade is reversed, with foreign business again accounting for the majority share.

Regarding the composition of the market, the role of government officials has been in steady decline, and there are fluctuations in majority share between business and leisure travellers, with both conference participants and tour groups important sectors of the total. If these are added to the respective categories, business tourism accounted for just over 42% and leisure tourism for 48% of the total at the end of 1999.

TRENDS AND CHARACTERISTICS

Several major trends that have affected the structure and the strategy of international hotel companies have been identified and will be discussed in detail. These are increasing consolidation and concentration; the increasing importance of branding and the impact of technology.

Industry Concentration and Consolidation

Chains in the hotel industry are getting larger, and becoming "mega chains" (consolidation), which also affects the total number of rooms under their control (concentration). Merger and acquisition activity is very prevalent in the international hotel industry. While a proportion of the rooms in the industry is coming under the control of a dwindling number of companies, this should be kept in perspective: two thirds of the total number of hotel rooms worldwide remains in independent (often SME) hands; the share of the chains is therefore by no means overwhelming.

Concentration in the industry can be measured by the share of the world market controlled by the major chains. Table 3 shows changes over the 1990s in the share of the worlds 200 largest chains.

Table 3. Percentage share of major hotel operators, 1990-1998

	1990	1995	1997	2000
Largest 200 hotel chains rooms ('000)	n.a	3856	4795	5660
Total number of rooms worldwide ('000)	12223	14116	14951	16300
% Share, top 200	n.a	27.3	32.1	34.5

Although the number of rooms worldwide has increased by 26% over the period, the share of the largest companies of the industry has risen throughout the period, and now stands at over one-third of the overall total. The 12 leading companies control 75% of the total number of bedrooms of the leading 50 chains, and have grown by 40% over the period 1995-200.

Despite being in the minority, it is the chains that set the pace and standards for the industry, as the vast majority of SMEs struggle to maintain profitability, or indeed survive. One of the major reasons behind the growth has been the chain's supply of consistent, standardised services and facilities that have been replicated throughout the world (also in developing countries) through standardised operating manuals and procedures. This avoidance of ambiguity, or the need for predictability, in such an experiential activity as staying in a hotel room, often in a foreign country, required the development of a brand name, which promises consistent delivery. Up to the 1980s, two of the most important strategic advantages which international hotel operators had over indigenous, national ones were a strong brand name and knowledge of international guests' requirements and preferences.

Industry consolidation may be seen from the growth in the size of companies that has principally occurred through merger and acquisition (M & A) activity. The international hotel sector is dominated by a relatively small number of companies. There are only nine corporations with over 100,000 rooms; seven of the nine originate in the US, which illustrates US dominance in the business. In the five-year period 1990-1994, the value of deals in the hotel sector was approximately US$22.4 billion. In the four years 1995-1998, this had risen to US$105.6 billion, increasing by a factor of almost five. Table 3.4 compares the top 10 hotel companies in 1975, with the top 10 twenty-five years later

Table 4. Comparison of Top 10 Hotel Companies, 1975-2000

	2000		1975	
Rank	Company	Rooms	Company	Rooms
1	Cendant Corp	548'641.00	Holiday Inn	274'969.00
2	Bass Hotels & Resorts	496'005.00	Sheraton Corp.	109'999.00

3	Marriott Int'l	404'792.00	Ramada Inns Inc	94'621.00
4	Accor	389'437.00	Hilton Hotel Corp.	61'621.00
5	Choice Hotels Int'l	349'392.00	Trust House Forte	60'705.00
6	Hilton Hotel corp	314'628.00	Howard Johnson	59'800.00
7	Starwood Hotels	215'207.00	Balkantourist	51'800.00
8	Carlson (Radisson/SAS)	119'710.00	Days Inn Inc.	37'983.00
9	Hyatt	87'969.00	Quality Inns Int'l	32'954.00
10	Sol Melia	82'733.00	TraveLodge Int'l	31'492.00
	Total	3'008'514.00		815'944.00

Source: "Service World", 1976, "Hotels", 2001.

The difference in size of the leading ten companies is very apparent. The number 1 hotel chain in 1975 (Holiday Inn) would have been in 7th position in 2000. Similarly, the number 2, (Sheraton) and number 3 (Ramada) would have been in 9 th and 10 th places respectively in 2000. Overall there are 3.7 times more rooms operated by the leading ten companies in 2000 than there were in 1975.

Only three of these companies, Bass, Accor, and Sol Melia, do not originate from North America. In Continental Europe, major companies in terms of market capitalization are: Accor (US$8.50 billion), Sol Melia (US$2.02 billion), Scandic (US$0.60 billion), Société du Louvre (US$0.73 billion), Movenpick (US$0.43 billion), and Groupe Partouche (US$0.43 billion). Significantly, even after some acquisition activity by Asian companies in the 1980s and early 1990s (with, for example, the purchase of ITT Sheraton and Kempinski Hotels by Asian corporations), Asian companies have no representation in the top 10, and are at the lower end of the top 20.

Importance of branding

Several advantages are attributed to brands, such as generating higher market share, price premiums from customers, providing higher investor returns, and establishing customer loyalty, thereby ensuring a more stable flow of earnings. In the hotel industry, this is especially important with regard to access to global reservation systems and purchasing power.

Brand penetration varies worldwide, however, with the highest ratio in the US, quoted at 70%. Other regions have far less branded hotel stock, with Asia and the Middle East recording 25%, South America 20%, and Europe only 15%, just over one-fifth of that of the USA. Europe especially has proved much more difficult for US chains to penetrate than was initially envisaged, owing to individual country characteristics (especially legal and financial restrictive practices), the size and age of properties and a lack of homogeneity which makes franchising (the main distribution method in the US), extremely difficult. Rates within Europe vary from over 20% in the UK and France to a miniscule 3% in Italy.

There are clear preferences for certain companies to expand by franchising (Cendant, Choice Hotels International and Bass Hotels and Resorts) and through management contracts (Marriott International, Société du Louvre and Accor) rather than by equity investment.

Until the mid-1990s, the names of the leading 20 hotel companies remained remarkably stable, with Holiday Inn, Sheraton and Intercontinental regularly appearing in the ranking. With increased concentration owing to merger and acquisition activity, however, these brands, although still in existence, appear as a multi-brand of the holding company, (Holiday Inn is owned by Bass Hotels and Resorts, which also owns Intercontinental; Sheraton is owned by Starwood, that also owns Westin; Days Inn is one of Cendant's brands, as is Howard Johnson). Cendant itself is a new company, having developed from the merger and acquisition of Hospitality Franchise Systems (HFS). HFS itself was only founded in 1990,with the fusing of several older, established brands such as Howard Johnson, Days Inn, Super 8 motels and Park Inns.

From the single brand name of the 1970s, all but one of the top ten chains had multiple brands by 1998. Analysts have suggested that within ten years the industry will be dominated by up to six "super-brands".

Impact of new technologies

In common with many other industrial segments, international hotel industry operations have been changed by new technologies. The most important changes have occurred through computer reservation systems and later global distribution systems that were originally developed by airlines (Galileo, Amadeus, Sabre), but were later adopted and personalised by major hotel companies. This has had a profound effect on the distribution of the hotel product. Secondly, the development of the internet has facilitated on-line reservations, which, although still very small in terms of the revenues generated (thought to be 2-5% of revenues for chain hotels), is nevertheless forecast to grow dramatically in the next few years.

In a recent survey, many European hotel chains were not technologically advanced enough fully to exploit the potential of the Internet through on-line booking, with accurate and timely price information on inventory available. Other effects of new technologies are evident in both customisation and purchasing.

Customisation

Marketing has evolved from mass marketing through segmentation (including psychographic and lifestyle segmentation) to database, customized, individual and finally cybermarketing. Customisation may be summed up by the memorable quote from Don Peppers and Martha Rogers, as "I know you. You tell me what you want. I make it. I remember next time ." Remembering in the hotel business has mainly been associated with the use of the Internet and the development of customer loyalty programmes. In practice, many of their features have been similar to traditional sales promotions, (including special offers and discounts). Table 5 shows the major company loyalty programmes by hotel companies.

Table 5 Major hotel companies' loyalty programmes

Company	Loyalty Programme	Number of Members(MN)
Marriott	Miles/Reward	13.1
Bass	Priority Club	8.9
Hilton	Hhonours	7
Starwood	Preferred Guest	5.1
Cendant	Various	5
Hyatt	Gold Passport	4.2
Radisson/Carlson	Gold Rewards	4.0
Choice	Guest privileges	1

Source: Peppers & Rogers, 2001

Some of the schemes are of vast proportions, with almost 30 million members in the three leading schemes (although the same person can be a member of all three). Information technology, therefore, allows the hotel companies to collect data on hotel guests in such areas as room rates paid, number of guests in party, frequency of stay, other services requested and the method of payment preferred. This information can then be used to tailor the company's offer to the needs of the customer. The Internet serves as the medium to identify and collect the personal information from the client, and then delivers the customized product.

In hotel companies this often translates into the offer of special features to members of the loyalty clubs who are given preferential treatment in areas such as on-line account management, awards redemption, stored profiles and quick reservations.

Industry examples of personalization include:

— Wyndham Hotels and Resorts (USA), since 1999, has built an online database of customer preferences including magazines, snacks, pillows, whether they wish to be near or far from the elevator and which floor they wish to be located on for all guests who have at least three stays per year.

— Fairmont Hotels & Resorts' (Canada) Loyalty programme is the "Fairmont Presidents Club;" benefits include instant check-in and check-out, room upgrades, free local calls, shoeshine, desired type of room, bed and pillow, distance of the room from the elevator, height of the room in the hotel etc. After a set number of nights, guests can be promoted to the next stages - gold and platinum that give higher room upgrades and awards.

— Marriott (USA) identified four major segments-business and leisure travelers, travel agents and event planners. Through Marriottcom, each of these four categories receive personalised services: a registered leisure traveler will for example receive links to an online vacation planner, a resort site and a golf site with details of the 150 golf courses at or near Marriott hotels etc.

Hotel companies wish to expand their Internet business for reasons of cost and potential. Regarding the former, it costs Bass $7.93 per net reservation booked through its call centre, compared to $3.20 online. Hyatt reports an even greater saving, with costs of $9 through their call centre and $3 online. Regarding the latter, potential for growth is considerable, when currently, on-line sales account for less than 5% of the total with many hotel companies. Mariott, however, forecast that online sales would triple in the two-year period 1999-2000.

Despite these examples, it has been argued that hotel companies have not so far made the fullest possible use of guest information, and loyalty schemes are often still used simply to reward levels of expenditure by guests rather than to analyse guest spending behaviour and patterns in depth, thereby missing the opportunity to personalize the offer fully and conduct one-to-one marketing.

Internet usage

Procurement

Business-to-business (B2B) purchasing via the Internet is increasing. Its advantage lies in bringing together of customers and suppliers. In the case of the hotel industry, this is a substantial business, taking into account all the purchases required e.g. furnishing, fittings and equipment (FFE), food and beverage supplies, service contracts, cleaning materials etc . One distinct trend is for several hotel corporations to form joint ventures to negotiate more favorable conditions via Internet purchasing. Marriott International, Hyatt and Club Corporation formed "Avendra" in May 2000.

It has been estimated that approximately US$ 7 billion per year could be recuperated from hospitality procurement by better use of the Internet, and by the end of 2000 up to US$30 billion of hospitality procurement will be on line.

On-line reservations

The Internet also provides major opportunities in saving costs and reducing complexity in distributing hotel rooms and marketing - especially customer relationship marketing (CRM). The major saving for a hotel company lies in reducing the number of intermediaries. Owing to the development of computerised reservation systems, global distribution systems and "switch" companies, the process of reserving a hotel room has become complex and expensive: if, for example a customer wishes to reserve a room through a travel agent the steps as shown in figure 2. occur:

Customer→ travel agent→ global distribution system→ switch→ central reservation system→ individual hotel.

Figure 2. Stages in the hotel room reservation process

As may be seen from figure 2, this is a complicated and time-consuming procedure, which has been estimated at US$7 plus 10% commission on the cost of the reservation. If "disintermediation" were to be applied by direct bookings through a hotel web site, reservations could be made for a small percentage of the cost of conventional methods.

Separation of hotel management from ownership

Several trends in relation to the ownership of hotel companies have had direct effects upon the structure of the industry, namely separation of the management of properties from ownership; divestment of non-core activities; vertical integration and rationalization.

Marriott was the first major chain to split into two distinct corporations, one dealing with the real-estate side of the business, the other with the management of the hotels. The divestment of businesses by major diversified companies has resulted in refocusing and a greater clarity of strategy; one example is Bass, which after several centuries of brewing beer and running public houses decided to focus solely on hotel developments.

To provide an overview of strategic developments in ownership throughout the industry over the past 15 years, an analysis was undertaken using the Ansoff matrix, which classifies activity into one of four areas: current product, current market; current product, new market; new product, current market; new products for new markets. This shows details of company ownership before and after acquisition. Several main trends may be identified:

1) Diversification -"new product, new market" - was principally a trend of diversification that started in the late 1980s, with companies from unrelated industries buying into the international hotel sector. This resulted in famous brands leaving the United States and going into Asian and European ownership. Intercontinental and Sheraton became Japanese-owned and Hilton International and Holiday Inn

became British-based. This trend ceased in the early 1990s, with the one major exception of the hostile Forte Group hotel takeover by the Granada Media Corporation in the mid 1990s.

2) Most of the activity occurred under the heading "current product for current market," with consolidation occurring on a significant scale as hotel corporations were merged and acquired. Horizontal integration has been important in the development of such chains as Hospitality. Franchising Systems (later Cendant), Accor, Hilton Hotels, and Bass Hotels and Resorts. This resulted in many of the chains coming back under the ownership of hospitality companies.

3) There has also been activity in the "current product, new markets" segment, as chains have sought to extend their geographic coverage (as in the acquisition by Bass of Inter Continental), or by attracting new market segments, by going either up or down-market, as in the case of the Marriott (mid-market) acquisition of Ritz-Carlton (deluxe market).

4) The fourth segment, "new products for current markets," has seen little innovatory activity. Such relatively new products as Formula One by Accor, boutique hotels (by independent and chain hotels), extended stay and suite hotels would be included in this category.

Expansion strategies of major international companies

Unlike manufacturing industries, hotels are obviously location-bound, and have only a certain number of market-entry modes available. The main considerations in determining the entry-mode selected is the level of equity investment demanded and the degree of control required over the operations (figure 3.)

The figure shows that there are non-equity strategies of management contracts and franchising (with lower levels of control and equity commitment) on one side, and acquisition of local operators and foreign direct investment on the other.

Strategic alliances and joint ventures may be seen to occupy the "middle ground," depending largely on the level of investment and other resources committed by the respective companies.

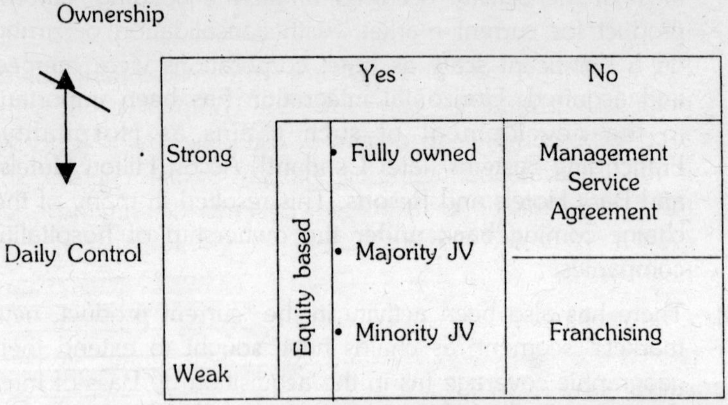

Figure 3. Relationship between control and ownership in the hotel industry

Control may occur at different levels, for example: (1) daily operational and quality control in each hotel property (2) control over the physical assets or the real estate (3) control over tacit expertise embedded in the routines of the firm (4) control over the codified assets such as the global reservation system and the firm's internationally recognized brand name.

The principal expansion modes (ranked according to level of control over the entrant) are as follows: Equity-involved strategy (high levels of resource commitment and control)

— *Long-term leasing*

This form of market entry is comparable to the full-equity method, as it commits the company to dedicate financial resources and assume control for a long period. Requiring as it does significant investment over time, long-term leases are given careful consideration by development directors, especially the corresponding risk and market desirability of the host country (for example in Germany where this is the preferred mode of entry), and has been used by companies in the past to secure sites in the best locations.

— *Foreign Direct Investment (FDI) and Mergers & Acquisitions (M & A)*

FDI in either a new property or a series of properties is a potentially profitable option for hotel companies, but with a correspondingly high degree of risk. The major concern is the political and economic stability in the country, in the light of the potential market in terms of demand and growth. As stated above, M & A has been a part of the hotel industry for many years.

There has been acceleration in merger and acquisition activity in recent years that may be attributed to both financial and marketing reasons. Regarding the former, investors from Asia and Europe benefited from changes in US tax legislation and the devaluation of the US dollar to invest in real estate and hospitality businesses in the United States. In the early 1990s, owing principally to the world economic recession, real estate prices dropped, allowing hotel companies with strong finances to acquire or merge.

Bass, for example, were able to pay $2.9 billion in cash to buy Intercontinental. There have been 13 transactions in these years of over £ I. billion, including the acquisition of ITT/ Sheraton ($8.23 billion), Forte ($ 5.9 billion), Doubletree ($ 4.64 billion) Intercontinental ($ 2.9 billion), and Stakis ($2.34 billion.) It is widely anticipated that this trend will increase in the short term, owing mainly to the need for companies to exploit economies of scale and scope.

Various splits of control and resource allocation

— Strategic alliance

A strategic alliance is a low-cost, limited-risk form of linkage or partnership between two or more companies. It allows multinationals to gain domestic market awareness and increase their network coverage. Forms of alliance range from relatively low-risk and low-visibility information sharing and referral of business to more overt joint branding. Strategic alliances are especially attractive to hotel companies operating in fragmented markets with small market share (e.g. Europe). Specifically, members from the alliance benefit from

— Greater market coverage

— Economies of scale and scope

— Increasing visibility of brand name

— Cross marketing

— Reduction in capital investment

— Complement each others strengths

— Market and reservation systems,

Recent examples in the industry include the agreements between the Dorint and Inter-Continental brands and Arabella & Sheraton. SAS and Radisson have worked together in their joint marketing and reservation systems for a number of years. Alliances may take the form of partnerships between hotel chains and food service providers, travel and transport agencies, computer and communication organizations, banks and other service organizations.

In June 2000, three of the largest European hotel chains launched an Internet alliance to achieve economies of scale and scope for online reservations and procurement.

Check and Dev distinguished three types of strategic alliance according to the time frame envisaged in the agreement, from short-term opportunistic to long-term strategic relationships. In theory, alliances allow firms to focus on their core strengths and offer a stronger product line with better market coverage. In practice, however, alliances are often characterized by high rates of failure, and the anticipated synergies are not as evident as first claimed.

— Joint venture

Joint ventures often take the form of partial equity investment and may involve the establishment of joint business projects with host countries or with regional partners. They may be comprised of companies with substantial financial resources, which in time lead to wholly owned subsidiaries with full ownership and control. By its nature the joint venture assumes a certain period of financial obligations and control, and has been used to good effect by multinationals in securing properties in key locations.

Non-equity-involved strategy (i.e. with low levels of resource commitment and control)

— Management contracts

The first hotel management contracts originated in the United States in the 1950s. The operation of the hotel property is separated from its ownership, and for some companies is a preferred type of expansion (e.g. Hilton International). Recently "pure management contracts" have become less common, as hotel developments are often linked in with other facilities (e.g. retailing, offices, medical etc.) The "package" therefore becomes more complex, with increasing equity participation from lenders (developers and host countries) demanded as a pre-requisite for the contract from the management contract company. There is a wide range in the duration of management contracts; Initially they were negotiated for periods of 20-25 years, but in the late 1980s and 1990s the trend was towards shorter terms.

A recent survey found contracts ranging across a span from 4-6 years to 20-24 years. Typically, operators will charge in the region of 3% of revenues as fees, plus incentive fees. As may be expected, large companies with high reputations can negotiate longer-term contracts with longer extensions; length of term correlates with brand name. Four Seasons, through their excellent brand name and reputation for high quality, have been known to command contract lengths of up to 70 years.

— Franchising

The International Franchise Association (IFA) defines franchising as "a continuing relationship in which the franchiser provides a licensed privilege to do business, plus assistance in organizing, training, merchandising, and management, in return for a consideration from the franchisee". The most common form of franchising is business format franchising, by which franchisers provide not only the brand name, but the complete business concept in a package, including the business plan, marketing strategy, management information systems and training. This format provides a channel of communication between franchiser and franchisee, in which the role of technology is becoming increasingly critical to the success of the latter: technology comes to play as important a role in creating competitive advantage in the hotel industry as in manufacturing.

There is an inherent problem in the expansion of franchising in the UK and mainland Europe, largely owing to the age of the hotel stock, which was developed before the era of standardisation and brand specification. Most hotels do not conform to the basic requirements behind major franchise brand demands. To date, only two hotel companies, Choice and Accor, have developed specifications that are flexible enough to accept the older European, mid-range hotel stock. Advantages for the franchiser include reduced capital investment, fast growth expansion, additional revenues/profits, and a potential for larger market share.

In the hotel industry, franchising was mainly used in the budget and mid-market sectors, but more recently has been effective even in the five-star bracket, e.g. through the franchise of the Four Seasons' Regent name to Carlson. In larger regions, master-franchising agreements are not uncommon, but the system is not without drawbacks. As shown in figure 3, franchising implies a lower level of control over quality, and conflicts are possible between the franchiser/franchisee over definition of territory, length of contract and compensation. The importance of cultural values should also not be underestimated. From a human viewpoint, the concept of franchising has been strongly criticized for the replication of the forms of the business that it produces: standardization, everywhere, is seen as socially unacceptable.

According to one authority, the enormous growth of chains resulted in:

> "...stultifying homogenisation of products and communities. They [the chains] destroy a sense of community by mass-producing environments that minimize personal contact. Moreover, since employees and managers are forced to operate inside a straightjacket of corporate regulations, franchisees have little chance to tailor their units to their own taste or local preferences."

This remains a minority view, however. Standardization, which is omnipresent, has often been seen as a relatively small cost to pay for the global development of the industry.

In addition to these market-entry strategies, hotel companies have adopted the hub-and-spoke development favoured by airlines. Typically, an up-market (four- or five-star) property will be developed in a gateway city. After this hub has been established, the spokes will be filled out by mid-market and budget properties in secondary and tertiary locations. At present, there are no cities in Eastern Central Europe that may be really considered "gateway". Prague, Warsaw and Budapest have been identified as having the potential to become gateway cities over the next few years, however. In this case, Mariott could use their Ritz Carlton brand in Warsaw, develop its standard Mariott hotels in, say, Krakow and Gdansk, and then use the Courtyard by Mariott in the tertiary locations such as Wroclaw and Opole,

Analysis of Expansion Strategies

As stated above, hotel companies often display a preference for non-equity forms of market expansion. Historically, this was mainly due to trends in the United States in the late 1960s and 1970s, when the value of real estate in the US rose faster than the rate of inflation. Investors therefore sought to purchase real estate as a hedge against inflation. This resulted in an increase in the supply of money for real-estate ventures, including hotels. Lenders increasingly insisted that hotel developers engage international franchise agreements as a condition for obtaining finance.

Simultaneously, tourism was beginning to gain recognition as an important economic factor in developing countries. Host governments therefore offered incentives to international companies for involvement in development projects. This increased to an all-time high of over 90% in the mid 1980s, but has subsequently stabilised to approximately two-thirds now. Among other causes, many hotel operators were obliged, in practice, to show some form of commitment through equity if they wished to gain major franchise contracts. As a result, the "pure" franchise is almost unknown at present.

The growth in non-equity arrangements is due to a convergence of factors, including a risk minimising strategy. Hotel operators pursuing international expansion may pursue both defensive and offensive strategies (i.e. avoiding losing home market share, and gaining significant new markets abroad). As competition in the home market increased, new external markets naturally beckoned. In the late 1970s and early 1980s, domestic markets were relatively small, and the hotel industry was characterised by a significantly high-risk structure: the risk derived from a dependence upon tourism factors such as seasonality and economic conditions within tourism-generating countries, including changes and fluctuations in disposable income and exchange rates. The capital-intensive nature of the industry also compounded the risk.

Furthermore, international hotel chains were seen to possess certain proprietary competitive advantages that would allow them to increase market penetration, including economies of scale, a more profound knowledge of international guest requirements, better-trained management and staff, better management and reservation systems, and strong brand name.

In addition to such factors, the hotel industry is well suited to non-equity forms of expansion owing to certain characteristics, including a willingness to transfer know-how, ease of codification of management systems, and the control that can be exerted over franchisees in the industry.

1) In contrast to other industries, there is indeed a willingness on the part of management contract and franchise companies to transfer their expertise and know-how to franchisees. A franchise may, in fact, be considered as much a marketing alliance as a form of transfer of corporate capability. The international hotel firm attempts to build up a global brand and reservation system, thereby satisfying final customers, and also appearing attractive to potential buyers in a highly competitive franchise market. The power of a franchise brand is a function of its size and the ability of the franchiser to drive demand into the hotels. It follows that the rate of expansion is a critical element in determining the success of a franchise operation.

   For the international operator, there are two critical factors to be considered when expanding internationally: firstly the franchiser's willingness and ability to supply the financial and human resources to the franchisees; secondly, the availability and quality of suitable licensees, franchisees, or partners in the target countries. Owing to the high fixed cost structure of the industry, there are attractive opportunities for marginal costing. With the high sunk costs of, for example, the development of global reservation systems, the cost of adding one more franchisee to the system is very small and any additional signings immediately increase profitability.

2) Owing to this codification of management and marketing practices, hotel corporations are able to provide comprehensive data and information, especially in relation to standard operating procedures. (SOPs), global reservations systems and likely financial projections. The problems associated with the "bounded rationality" of franchisees are reduced. The information also assists

franchisees in their attempts to capitalize on the value of the brand name.

3) The franchiser continues to exert control over the franchisee, however, through the right to use the brand name and reservation systems. Franchising hotel corporations are not afraid of creating competitors or of dissenting behaviour from contractees, as the threat of the loss of the system privileges is strong.

9

Future of Hotel Organisation

No hotel company operating today can be unaware of the swift pace of global change and its impact on every facet of the hospitality industry. With just five years left in this century, we can expect change to be the only constant. New business practices are evolving virtually as fast as our technologies, while resistance to change has become one of the primary causes of business failure.

The specter of constant change raises fundamental questions as to the creation of shareholder wealth in a capital-constrained, highly competitive environment. How will hotel organizations build shareholder wealth, and what key drivers will result in success? What future products and services will be essential in a technology-driven, global environment marked by rising customer expectations? Moreover, what alternative approaches and skills must organizations develop to ensure market success?

The future success of hotel organizations will be driven in large part by the ability to foresee — and capitalize — on change. Beyond this truism, however, there is an urgent need to identify what will be required in the competitive environment of the future with its intense focus on serving customer needs. The hospitality industry — as is the case with business generally

— is subject to deep currents of change set in motion as economic and social systems shaped in the industrial era evolve to a knowledge-based era driven by technology advances.

In this period of global transition, it behooves hotel organizations to examine the key factors that will not only define success, but the ability to survive in coming years. Many of these issues were spotlighted in a global study undertaken by the Economist Intelligence Unit and co-sponsored by Arthur Andersen. The Successful Corporation of the Year 2000 surveyed more than 600 senior executives around the world. Its mission was to identify the characteristics needed to lead successful businesses in the next century. These executives offer a number of compelling messages regarding the key success factors of the future.

Customers will have the strongest influence on the corporation in the year 2000. Indeed, these executives believe that the customer will be "king" in the new century. Exceptional leadership was by far the attribute most frequently cited by CEOs and senior executives; the consensus appears to be that successful companies in the year 2000 will be led by corporate visionaries. A strategic planning focus is not only essential, but must embody a concept of planning for the future that anticipates change, rather than being based exclusively on historical models. In structuring organizations for the future, companies must build management capabilities to deal with one of the most critical challenges — diversity in the marketplace.

Employing information technology to drive business success in this information-driven era is not only the path of least resistance, but vital to virtually every aspect of operations. The study's top-line results create an excellent backdrop to address key issues confronting the hotel industry, and what factors will lead to membership — rather than rejection — in the elite club of the world's corporate success stories in the year 2000.

Recognizing that the hotel industry has a somewhat split personality reflecting the inherent conflicts between its real estate and operational aspects, it is important to understand the industry's real estate origins and how they are shaping the challenges ahead. The origins of the industry's real estate persona are embodied in the classic theory of location - "if we build it, they will come." As a result of this "building" complex, the industry has tended to have a real estate and asset orientation, rather than a customer focus. From the hotel company perspective — especially that of the brand-oriented "chain" — the varied interests of a diffused property ownership group can be quite different than the singular interest of the chain that operates and markets the properties.

Even when the ownership of geographically dispersed hotel properties is controlled by a single hotel organization, the financial structuring tends to be property-specific. Corporate financial strategies are frequently subjugated to the needs of the last property deal brought into the company's fold. Each property in a so-called "chain," is frequently the subject of a unique and distinctive ownership and financial structure. This phenomenon — quite common in the real estate sector, but unusual for business enterprises generally — makes for elusive economies of scale in the structuring and financing of property-driven expansion.

Collectively, these factors have produced low comparative returns in real estate, although criticism leveled at commercial real estate returns is somewhat less germane to the hotel industry, where management and franchise fees can produce high returns for those companies where property ownership is held by third parties. Nevertheless, average returns in the real estate industry in the United States, for example, are just over 10 percent, compared to small company stock returns at almost 20 percent and large company stocks at an average of about 15 percent.

While hotel chains have adopted traditional corporate frameworks, there are a number of predominantly real estate-driven, family-owned businesses in the hotel sector that continue to operate as relatively unstructured organizations. In meeting the future, these businesses will need plans, people and processes in order to establish viable corporate forms that can compete in tomorrow's marketplace and capitalize upon its opportunities.

In today's changed environment, the hotel organization must deal with a number of new realities. Investors in our industry are no longer satisfied with long-term capital appreciation and psychic income that heretofore were often the justification for otherwise seemingly uneconomic investments in hotel property or, indeed, hotel chains. The first reality is that there is a very specific and identifiable relationship between bottom line performance and value. Improvements in business operations raises values. It is not surprising, therefore, that the new owners attracted to this industry in recent years have new sets of demands. This transformation from an old-guard group of investors and owners to income and return-driven newcomers has meant that the once "quiet enjoyment" of operators in their management of hotels for third parties is being disturbed, interrupted and overturned.

These challenges all take place in an environment where capital has become extremely selective in markets that have little stability. A global shortage of capital will not remain a short-term problem, and future hotel organizations must have a stronger alignment to capital providers — a critical "customer" group. Hotel companies will need to compete by offering better returns and performance than in the past. Hotel chains have found their development timetables quashed in recent years, making it difficult to achieve goals of critical mass often required to improve performance. These factors are driving consolidation in the brand "sweepstakes." Capital markets, therefore, continue to favor well-established companies, a reality that must drive entrepreneurial organizations to meet the future now by planning for an evolving corporate context in which to operate.

Future hospitality Manager and the Legal Environment

Today more than ever before, hospitality managers must be multitalented individuals. In addition to knowledge of their own designated area of expertise, such as food and beverage, marketing, accounting, or rooms management, hospitality managers are often called upon to assume specialized roles, such as employee counselor, interior designer, facility engineer, or computer systems analyst. Given the complexity of the modern business world, it is simply a fact that the skill level required for success today in this field is greater than it was in the past.

Hospitality management has always been a challenging profession. Whether in a casino, a school lunch program, a five-star hotel, a sports stadium concession program, or myriad other environments, hospitality managers are required to have a breadth of skill not found in many other areas of management. Hospitality managers are in charge of securing raw materials, and producing a product or service and selling it-all under the same roof. This makes them very different from their manufacturing counterparts, (who are in charge of product production only), and their retail counterparts, (who sell, but do not manufacture, the product).

Perhaps most important, the hospitality manager has direct contact with guests, the ultimate end users of the products and services supplied by the industry. Additionally, hospitality managers are called upon frequently to make decisions that will, in one manner or another, impact the legal standing of their employers. Robert James, founder of one of the largest hotel contract management companies in the United States once estimated that 60-70 percent of the decisions he made on a daily basis involved some type of legal dimension. This is not to say that hospitality managers need to be attorneys. They do not. However, the decisions that they make may or may not increase their organization's chances of needing the services of an attorney.

Consider the situation where a hospitality manager is informed that a guest has slipped and fallen in an area of the dining room containing a salad bar. It appears that the guest had been serving himself and slipped on a piece of lettuce dropped by a previous guest. Was this a simple accident? Could it have been prevented? Is the restaurant responsible? What medical attention, if any, should the manager be prepared to provide? What if the injuries are severe? Should the restaurant be held responsible? Can the restaurant manager be held personally responsible? Most important, what should the manager actually do when the incident is brought to his or her attention? What, if anything, should the employees do? Who is responsible if the employees were not trained in what to do?

From this example, it is clear that the hospitality manager is in a position to profoundly influence the legal position of the operation. Day after day, in hundreds of situations, the actions of hospitality managers will influence the likelihood of the business or the manager becoming the subject of litigation. There is a unique body of law relating to the food service, travel, and lodging industries. These laws have developed over time as society and the courts have sought to define the relationship between the individual or business serving as the host and the individual who is the guest.

Jack P. Jefferies, who served for more than 20 years as legal counsel for the American Hotel and Motel Association (AH&MA), has stated that: "Over 135,000 new federal and state laws are issued annually, as well as hundreds of thousands of federal and state administrative rules." With this much change in the law, some believe that the topic is too complex to learn in an introductory course or from one book. In addition, they would argue that because the law is constantly changing, even if an individual learned the law today, his or her knowledge would be out of date in a very short time.

While these positions are understandable, they argue for, not against, the future hospitality manager's study of legal management. While the law is indeed complex, certain basic principles and procedures can be established that will minimize a manager's chances of encountering legal difficulty. Because that is true, it is less important to know, for example, the specific rules of food safety in every city than it is to know the basic principles of serving safe food. No one, not even the best lawyer, can be expected to know everything about every area of the law. In the same way, hospitality managers are not required to have a comprehensive knowledge of every law or lawsuit that impacts their industry. What they must know, however, is how to effectively manage their legal environment. To begin this journey, it is important that they:

1. Know the historical origins of the law.
2. Recognize that laws have an evolutionary nature, based on changes in society.
3. Understand how to use a philosophy of preventative management to manage the legal environment and minimize the chances of litigation.

Common law and civil law are the two major systems of law in place in the western world. Common law is the body of law that descended from Great Britain and is used in the United States and most countries in the British Commonwealth. Civil law descended from the Roman Empire and is used by most Western European countries, as well as Latin America, Asia, and Africa. While both legal systems certainly defy oversimplification, it can generally be said that common law comes from reviewing past litigation that has been decided by the courts. It is greatly interested in precedent, or what has been decided in previous court cases with similar situations or facts.

In civil law, decisions evolve based on written laws or codes. Judges in civil law feel less bound by what others have decided before them and more compelled by the law as it has been established by government bodies. Given the nearness of countries within Europe, and the influence of the British Empire, it is no wonder that these two great legal systems frequently operated in close proximity, thus often blurring their distinctions. Interestingly, the term civil law is actually used in the common law system to refer to private law (or private disputes), as opposed to public or criminal matters.

Common law developed in England following the Norman Conquest. Essentially, the purpose of the common law was to interpret and enforce rules related to the granting of land by the British monarchy to those subjects deemed worthy of such land grants. The barons who received this land would often grant parts of it to those they felt deserving. The courts that were created at this time were charged with overseeing the peaceful resolution of disputes regarding land, inheritance, marriage, and other issues related to land grants.

The Evolutionary Nature of Common Law

It should come as no surprise that a rapidly changing society will often revise its rules of conduct and responsibility. This is true in society as a whole and in how society views the hospitality industry. In the United States of the 1850s, obviously, one would not have been expected to find a law requiring a certain number of automobile parking spaces to be designated for disabled individuals seeking to enjoy an evening meal at the town's finest restaurant, because the world in that era contained neither the automobile nor the inclination of society to grant special parking privileges to those who were disabled.

In today's society, we have both. What changed? First, the physical world changed. We now have automobiles, along with the necessity of parking them. More significant, however, is the fact that society's view of how disabled individuals should be treated has changed. Parking ordinances today require designated "disabled" parking spaces, generally located close to the main entrances of buildings to ensure easy access. Not only is it good business to have such spaces, current laws mandate that the hospitality manager provide them.

In this case, parking requirements grew out of a law created at the federal government level. The law is called the Americans with Disabilities Act (ADA). Laws in the United States may be enacted at the federal, state, and local levels. At each of these levels, the laws reflect the changing desires of the citizens and their elected officials. Because society includes members who operate hospitality facilities, hospitality-related laws created and modified by society impact those who work in the hospitality industry.

Future hospitality managers will encounter laws that do not currently exist. How then can they be expected to operate their facilities in full compliance with the law throughout their career? Just as important, how can they be expected to manage these facilities in a way that will minimize their chances of doing something illegal? The answer is not found in an attempt to monitor every legislative body empowered to enact law. The answer is to operate hospitality facilities in a way that combines preventative legal management with sound ethical behavior.

In the medical field, it is widely agreed that it is better to prevent a serious illness beforehand than to treat it after the fact. For example, doctors would advise that it is preferable to prevent a heart attack through proper diet, exercise, and the cessation of smoking than to perform a bypass operation on a patient after a heart attack has occurred. In the case of prevention, the doctor advises the patient, but it is, in large measure, up to the patient to put into practice the recommendations of the physician.

In a similar vein, it is far better for hospitality managers to operate their facilities in a way that minimizes the risk of litigation, rather than in a manner that exposes their operations to the threat of litigation. As noted, the law is not static; in fact, it changes frequently. Managers must stay abreast of these changes so that ultimately, on a daily basis, they integrate their acquired knowledge and awareness of the law into a personal style of management and decision making. The STEM acronym was coined as an easy way to remember the steps in a decision-making process that can assist managers in getting started. It stands for: select, teach, educate, and manage. It is presented here as a way of beginning to "STEM" the tide of litigation.

On any given day, the General Manager of a hotel or restaurant in the United States will make decisions about hiring, firing, and/or providing benefits to employees. Other daily tasks might include approving a meeting space contract for a major event to be held on the property, an event that involves the service of alcohol. Decisions regarding if and when to add a lifeguard to the pool area, whether to subcontract parking services to a local valet company, and even the uniform requirements of staff will all be made by the manager. All of these seemingly independent decisions have a significant common denominator: their legal implication.

Whether it is opening a restaurant, operating a country club, or hiring a housekeeper, hospitality managers must be aware of the legal implications of each and every decision they make. It is of vital importance that managers resolve to be fair, to operate within the law, and to manage preventatively. On occasions when they do not, and a lawsuit results, the courts may hold managers liable for their inattentiveness. This philosophy of preventative management becomes even more important when one considers that a great many litigation matters encountered by hospitality operators have a common denominator: a poorly prepared employee.

Injuries and the resulting damages, whether financial, physical, or mental, are usually a consequence of an employee who has not been sufficiently taught to perform his or her duties. He or she may make an omission, such as not cleaning up a spill near a salad bar, or he or she may pursue an activity outside the scope of his or her duties, such as sexual harassment or arguing with a customer. The recent increasing number of lawsuits is not caused solely by employees, of course. The legal system and some attorneys certainly share the blame. Managers, however, bear most of the responsibility for what has been occurring.

When an employee makes a mistake, often it is the result of management error. Either the wrong person was hired for the job, the duties for the job were not effectively communicated to the employee, the employee was not properly trained, or the employee was not effectively supervised or motivated to do the job properly. To create an environment conducive to motivation, you must first establish trust and respect. When managers make a commitment to employees or guests, they must follow through. They also must be willing to accept responsibility for their mistakes, and to apologize for them when appropriate. Managers must set an example: If a manager asks employees to be on time, then the manager must also be on time; if managers expect employees to pay for food, beverages, and services, then they must also pay for food, beverages and services. In current parlance, managers must walk the talk!

Finally, all of the planning, organizing, controlling, and motivating in the world will not help if management cannot effectively communicate its vision and plan to the employees who will carry out that vision. The ability to communicate with skill and grace is a critical component of being a successful manager. Today's culturally diverse workforce will require diverse motivating techniques. Remember that different people are motivated by different incentives. Money is a perfect example. To some, it is a strong motivating factor; others would prefer more time off instead of additional pay. Managers must know their employees, and determine, by asking them if need be, what would motivate them, both as individuals and as a work-team. Examples of possible motivational efforts include:

— A sales contest with a significant prize.

— A parking space with recognition for the employee of the month.

— A 50 percent discount on meals at the restaurant.

— A card on their birthday.

— A written "pat on the back" for a job well done.

— Taking the time to ask them how their day was.

— Involving employees in setting goals.

— Seeking employee input in developing work schedules.

— Listening to their concerns.

All of the above and others are the types of activities a manager should undertake to build the trust and loyalty of employees. If a consensus can be reached on what to do and how to do it, the motivating task becomes much easier. The goal of STEM is to reduce employee mistakes. By continually encouraging and rewarding good performance, managers can create an environment that will, in fact, reduce the number of times employees make mistakes. Remember that even though a goal may not be reached, the efforts of the individual or group still may merit praise.

In other words, managers should try to catch their employees doing something right instead of trying to catch them doing something wrong. It is not possible to manage effectively while sitting behind the desk. Effective managers know that "management by walking around" is alive and well, particularly in a service industry such as hospitality. Of course, an important part of managing is the ability to motivate employees. As much as managers would like all employees to come to the job every day brimming with enthusiasm, the fact is, too often, just the opposite is true.

A significant number of employees may dislike coming to work, dislike their jobs, their situation in life, and much more. They must be motivated to perform at the level management has targeted in order to exceed management's own expectations, and, more important, those of the guest. To recap the STEM process: Select the right employee for the right job; teach employees while creating a training trail; educate management; and motivate staff in a positive and nurturing manner. All these efforts will help foster loyalty and goodwill while reducing the likelihood of litigation.

Ethics and the Law

It may not always be clear whether a course of action is illegal or simply wrong. Put another way, an activity may be legal, but still the wrong thing to do. As a future hospitality manager who seeks to manage his or her legal environment and that of other employees, it is important that you be able to make this distinction. Ethics refers to the behavior of an individual toward another individual or group. Ethical behavior refers to behavior that is considered "right" or the "right thing to do."

Consistently choosing ethical behavior over behavior that is not ethical will go a long way toward avoiding legal difficulty. This is true because hospitality managers often will not know what the law requires in a given situation. In cases of litigation, juries may have to make determinations of whether a manager's actions were intentionally ethical or unethical. How juries and judges decide these questions may well determine their view of a manager's liability for an action or inaction.

While it may sometimes be difficult to determine precisely what constitutes ethical behavior, the following seven guidelines can prove to be very useful when evaluating a possible course of action:

1. Is it legal?
2. Does it hurt anyone?
3. Is it fair?
4. Am I being honest?
5. Would I care if it happened to me?
6. Would I publicize my action?
7. What if everyone did it?

Using the seven ethical guidelines just listed, a manager could evaluate whether he or she should reduce the bride and groom's final bill by the difference in selling price of the two champagnes. How an individual determines what constitutes ethical behavior may be influenced by his or her cultural background, religious views, professional training, and personal moral code. A complete example of the way someone would actually use the seven ethical guidelines is demonstrated in the following hypothetical situation:

An Ethical Situation Assume that you are the Food and Beverage Director of a large hotel. You are planning for your New Year's Eve gala, and require a large amount of wine and champagne. You conduct a competitive bidding process with the purveyors in your area, and, based upon quality and price, you place a very large order (in excess of $20,000) with a single purveyor. One week later, you receive a case of very expensive champagne, delivered to your home with a nice note from the purveyor's representative stating how much they appreciated the order and that they are really looking forward to doing business with you in the years ahead. What do you do with the champagne? Ethical Analysis Your first thought may be the most obvious one; that is, you drink it. But hopefully, you will first ask yourself the seven questions of the ethical decision-making process.

1. Is it legal?

From your perspective, it may not be illegal for you to accept a case of champagne. However, there could be liquor laws in your state that prohibit the purveyor from gifting that amount of alcoholic beverage. You must also consider whether it is permissible within the guidelines established by the company for which you work. Many companies have established gift acceptance policies that limit the value of the gifts that employees are eligible to accept. In this case, violation of a stated or written company policy may subject you to disciplinary action or even the termination of your employment. Accordingly, you need to be extremely familiar with the ethics policy that has been adopted by the company you are working for. Assuming that it does not violate a law and/or company policy, go to question 2.

2. Does it hurt anyone?

Well, it probably would not hurt you, unless you drank all of the champagne at once, but, realistically, are you really going to be fair and objective when you evaluate next year's bids, or is your mind going to be thinking back to the case of champagne that you received? Assuming that you do not think that it is hurting anyone, go on to question 3.

3. Is it fair?

Before answering this question you have to recognize who the stakeholders are in this particular situation. How might others in your company feel about the gift you received? After all, you agreed to work for this firm at a set salary. If benefits are gained because of decisions you make while on duty, should those benefits accrue to the business or to you? Assuming that you have decided that it is fair for you to keep the champagne, go to question number 4.

4. Am I being honest?

This question gives you the opportunity to second-guess yourself when you are answering questions 2 and 3. Do you really believe that you can remain objective in the purchasing aspect of your job, and continue to seek out the best quality for the best price, knowing that one of the purveyors rewarded you handsomely for last year's choice and may be inclined to do so again?

5. Would I care if it happened to me?

If you owned the company you work for, and you knew that one of the managers you had hired was given a gift of this magnitude from a vendor, would you question the objectivity of that manager? Would you like to see all of your managers receive such gifts? Would you be concerned if they did?

6. Would I publicize my action?

If you have trouble remembering the other questions, try to remember this one. Would you choose to keep the champagne if you knew that tomorrow morning the headlines of your city newspaper would read: "Food and Beverage Director of Local Hotel Gets Case of Champagne After Placing Large Order with Purveyor"? Your General Manager would see it, other employees would see it, all of the other purveyors that you are going do business with would see it, and even potential future employers would see it.

7. What if everyone did it?

If you justify your choice of keeping the champagne, consider: does this process ever stop? What would happen if the Executive Housekeeper had a bed delivered to her home every time she ordered new bedding for the hotel? What would happen if every time she ordered new washers and dryers, she received a matching set at home?

What are some of the realistic alternatives to keeping the champagne?

— Return it to the purveyor with a nice note telling them how much you appreciate it, but that your company policy will not allow you to accept it.

— Turn the gift over to the General Manager to be placed into the normal liquor inventory (assuming that the law will allow it to be used as such).

— Donate it to the employee Christmas party.

Use the seven questions to evaluate each of these three courses of action. Some hospitality managers feel it is important to set their ethical beliefs down in a "code of ethics." Ethical behavior, however, is always important to the successful guidance of responsible and profitable hospitality organizations.

SUCCESS FACTORS FOR THE FUTURE

Within this broad context, hotel developers, owners and management companies will all need to develop new strategies, skills and processes that look forward to the competitive demands of the future. These ultimately must address issues related to vision and planning, as well as organizational skill sets and processes to attract and retain customers. To stake a claim in the future, current business practices should be examined in light of what can be expected to be the key success factors in the year 2000.

As the information age produces greater worldwide integration of business activities, a global knowledge base will become invaluable. Success in local and regional hotel markets will be shaped decisively by a global business environment that defines capital movement, customer expectations and applications of new technologies.

If the customer is "king" in the 21st century, hotel organizations will be best served by focusing less on their hotel assets as measures of success, and more on their customers. This involves a fundamental shift in viewing the real estate asset as the wealth creator — to the customer as the key to building shareholder wealth. A customer focus must imbue business decisions at all levels of developing and operating a hotel organization. Pursuing such a course will inevitably impact shareholder wealth. To accomplish this, however, customers need to participate in the product development process.

Fully realizing a customer focus in the industry poses a significant challenge. The hotel industry must confront problems due to conflicts between operational needs and real estate goals. Quite simply, an operator must remain customer-focused, but the short-term strategies to meet these needs may be inconsistent with the long-term objectives of property owners. Balancing those goals will be essential. A customer. focus implies a significant shift in what drives hotel development — placing primary emphasis on the customer with the locations to follow.

Nevertheless, a hotel organization with its large investment in fixed assets — the real estate — can never be as nimble as a consumer products company in adjusting products and services to match rapid shifts in the marketplace. The Japanese taught us that the concept of "zero defects" in products and services can yield tremendous benefits. But today an even more rigorous standard dominates — quality that surprises. In practical terms, the hotel industry finds it extremely difficult to meet the standard of zero defects in service.

Hotel services are based primarily on people, not computers or other equipment. Quality that surprises takes the concept of zero defects a step further. Yesterday's surprising product or service is today's status quo. Twenty years ago, a business executive did not expect a consistent and predictable level of service wherever he or she traveled in the world. Today that is a standard — not the exception — as is the expectation for sophisticated technology in hotel rooms to support business needs. With customer discrimination so acute, it is not surprising that brand loyalty is a diminishing "commodity" in the hotel industry.

Find the ingredients for visionary leadership. Today's hotel organizations need to recognize the need for visionary leadership. The ability to forecast the future — to anticipate change rather than react to it — will be one of the single greatest determinants of market dominance in the years ahead. The old "command and control" model of leadership is giving way to a focus on leadership in ideas, information, inspiration, vision and teamwork. Warren Bennis, an authority on leadership in the U.S., puts it this way — "For the most part, failing organizations tend to be over-managed and under-led. The leader sets the tone for the moral character, the vision, the corporate culture and the fiber of the institution."

While visionary leadership is essential, it must be linked to business operations and foster a risk bias, rather than a procedural bias. This will allow the organization to stretch and, in turn, change. And it must be shared by empowered professionals and staff throughout the organization, including those who meet the customer. Overcoming the resistance to change can be a daunting task, particularly in large hotel organizations in which diversions from the status quo may threaten established management lines. It suggests that a culture based on conformance may need to be replaced by an emphasis on flexibility, learning and cooperation.

Management competencies will need to be aligned in order to achieve the desired result. For many organizations, this may mean a shift from traditional hierarchies typical of companies in an industrial era to a flatter organization with a more transparent interface between leadership, organizational functions and employees. For many of the industry's leaders, vision is driven by the strategic planning process, a function which has become critical for success.

Strategic planning, however, has at times been a step-child in the hotel industry, and it is often the first to be cut when organizations are downsized. It is clearly in transition. There also has been a tendency to decentralize and simplify this function — both actions of potential benefit. Strategic planning must be led by the top people in the organization — the CEO and COO. On the other hand, it should be close enough to the customer to ensure that planners can "listen" to and be influenced by customer needs.

Beyond the ability to envision the future, core management capabilities will make the difference — they are essential. A clear vision without the management skills to support it cannot be a recipe for success. First and foremost, hotel management must have strategic development skills and the ability to integrate complex factors affecting success. Market volatility has become the norm, in part caused by the rapidly changing tastes of customers. Customers are increasingly approaching the hotel industry with widely different social, economic and political backgrounds, to say nothing of employees. Being able to deal with these diversities in a positive and constructive fashion that capitalizes on the differences, rather than working to find ways to mitigate them, is the clear path for successful management in the future.

The organization will also need to be imbued with a sense of entrepreneurship that reacts proactively to the market's diversity. Traditional organizations that follow well-documented rules must give way to leaders who can balance a sense of discipline with that of flexibility. Talent and resources must be marshaled and leveraged. In an industry with high fixed costs and labor intensity, the concept of leverage in the hotel business is an all-important one. Improving labor productivity through technology must be a goal for today's forward-looking hotel organization. Management must also be able to narrow the gap between the employer and the employee, forcing a flatter organization in the process. This will put management closer to the customer and speed the two-way communication process up and down the organization.

The traditional role of information technology (IT) as a back office support for accounting and bookkeeping has clearly moved front and center stage. IT today influences all aspects of business from corporate strategies to organizational structure — and from the very business processes it is designed to support to performance measurement. In a world where the customer is "king," IT must also deliver in two critical areas: sales and marketing and customer service.

Technology was once viewed as a way to reduce costs by replacing people. That attitude has been firmly supplanted by one that seeks IT support for the creative work that all organizations must pursue. IT must allow organizations to react more speedily to market needs and, of course, produce the fulfillment of customer demands both quickly and accurately. To do this IT must operate on a decentralized basis. IT delivers, but it has to be the right information to the right people, and it needs to be done on a timely basis.

Bibliography

TEXT BOOK REFERENCING

Abbot, P. *Front Office : Procedures, Special Skills and Management,* Oxford : Butterworth-Heinemann, 1991.

Arthur, R., and D. Gladwell. *The Hotel Assistant Manager.* 3rd ed. London : Barrie & Rockliff, 1975.

Baird, C., and L. Carla. *Front Office Assignments.* London : Cassell, 1992.

Bardi, James A. *Hotel Front Office Management.* New York : Van Nostrand Reinhold, 1990.

Bardi, J.A. Hotel Front Office Management. New Jersey : John Wiley & Sons, 2003.

Boomer, Lucius. *Hotel Management.* New York : Harper & Row, 1938.

Braham, Bruce. *Computer Systems in Hotel and Catering Industry,* London : Cassell Education Ltd., 1988.

Borsenik, Frank. *The Management of Maintenance and Engineering in the Hospitlity Industry.* 3rd ed. New York. John wiley & Sons, 1992.

Browning Marjorie. *Night Audit Procedure.* Columbus The Christopher Inn, March I, 1969.

Burstein, Harvey. *Hotal Security Management.* New Your: Praeger Publishers, 1975.

Buzby, Walter J. *Hotel and Motel Security Management.* Los Angeles : Secutity World Publishing Co., 1976.

Chandler, Raymond. *Trouble Is My Business.* New Youk : Ballantine Books, 1972.

Collins, Galen. *Hospitality Information Technology: Learning How to Use It.* Dubuque, IA: Hunt Publishing, 1992.

Dukas Peter. *Hotel Front Office Management and Operations*. 3rd ed. dubuque, IA:Wm C. Brown, 1970.

Foster, Dennis. *Rooms at the Inn: Front Office Operations and Administration.* Lake Forest, IL: Glencoe, 1992.

Front Office and *Reservations*. Burlingame, CA: Hyatt Corporation, 1978.

Gomes, Albert J. *Hospitality in Transition*. Houston: Pannell Kerr Forster, 1985.

Kasavana, Michael L. and Richard M. Brooks. *Managing Front Office Operations.* EI of AH&LA, 2001.

Kasavana, Michael L. and Richard M. Brooks. *Managing Front Office Operations.* EI of AH&LA, 2002.

Kasavana, Michael L. and Richard M. Brooks. *Managing Computers in the Hospitaity Industry*. 2nd ed. El of AH&LA, 1992.

Lundberg, Donald. *The Hotel and Restaurant Business.* 5th ed. New York: Van Nostrand Reinhold, 1989.

Orkin, E. Boosting Your Bottom Line with Yield Management, Cornell Hotel and Restaurnat Administration Quarterly 30, no. 3 (November 1989).

Property Management and Point of Sale Systems: Guide to Selection. New York. EI of AH&LA (date unknown).

Rutherford, Denney G. *Hotel Management and Operatinons*. New York: Van Nostrand Reinhold, 1989.

Starting and Managing a Small Motel. Small Business Administra tion, Washington D.C.

The State of Technology in the Lodging Industry. New York. El of AH&LA, 1980.

Troy *Keys to Security.*

Yield Management Systems, (YMSI, L.C.S.), Chicago, U.S.A.

Vallen, G.K. & Vallen, J.J. *Check-in Check-out*. Prentice-Hall, New Jersey, U.S.A., 1996.

Valen, G.K. & Abbey, J.A. *The Art and Science of Hospitality Management*. EI of AH&LA, 1987.

ELECTRONIC REFERENCING

http: // www.acfchefs.org

http: // www.ahma.com

http: // www.astanet.com

http: // www.csshotelssysms.com

http: // www.cssmain.com

http: // www.ei-ahla.org

http: // www.executech.com

http: // www.fhrai.com

http: // www.hitecshow.org

http: // www.hitis.com

http: // www.hospitalitynet.nl.com

http: // www.hostgroup.com

http: // www.hotelchoice.com

http: // www.hrnai.com

http: // www.invest.com

http: // www.lodginsystems.com

http: // www.micros.com

http: // www.newsoft.com

http: // www.penton.com

http: // www.resortdata.com

Index